Rivers
of the
United States

Rivers of the United States

VOLUME V
PART A: THE COLORADO RIVER

Ruth Patrick

FRANCIS BOYER CHAIR OF LIMNOLOGY
THE ACADEMY OF NATURAL SCIENCES OF PHILADELPHIA

JOHN WILEY & SONS, INC.

New York • Chichester • Weinheim • Brisbane • Singapore • Toronto

This book is printed on acid-free paper. ∞

Copyright © 2000 by John Wiley & Sons, Inc. All rights reserved.

Published simultaneously in Canada.

This publication is designed to provide accurate and authoritative information in regard to the subject matter covered. It is sold with the understanding that the publisher is not engaged in rendering professional services. If professional advice or other expert assistance is required, the services of a competent professional person should be sought.

Library of Congress Cataloging-in-Publication Data:

ISBN 0-471-30345-3 (v. 1)
ISBN 0-471-10752-2 (v. 2)
ISBN 0-471-30346-1 (v. 3)
ISBN 0-471-19741-6 (v. 4, Pt. A)
ISBN 0-471-19742-4 (v. 4, Pt. B)
ISBN 0-471-30347-X (v. 4, set)
ISBN 0-471-30348-8 (v. 5, Pt. A)

Printed in the United States of America.

10 9 8 7 6 5 4 3 2 1

Contents

Preface

This volume is the fifth in a series of books concerned with rivers of the United States. In the first part, the general structure of the riverine system and ecosystems in the main stem of the Colorado River and selected tributaries are described. In the second part, salinity and alkalinity problems are discussed and the laws, regulations, and some of the politics are set forth.

The Colorado River is often described as the most manipulated river in the world. It is a relatively small river compared to other riverine systems in the United States and throughout the world. Nevertheless, it is a very important river because it furnishes the water for much of the states of Colorado, Montana, Utah, Nevada, New Mexico, and Arizona and supplies water to certain areas of California.

The Colorado River arises on the western slope of the continental divide in Colorado, and shortly after its origin, picks up the Green River, which arises in Montana. It flows southward, picking up the San Juan River in Colorado and the Gila River in Arizona. The Colorado has many other tributaries, but these are the principal ones. Due to the great amount of manipulation of the river, mainly for agriculture and power but also for municipal purposes, the river is severely altered. In very few areas do ecosystems of aquatic life exist that approach those found in natural conditions. Since early days the river has been altered for irrigation for agriculture purposes. More recently, dams have been built to generate electric power. The largest of these is the Hoover Dam.

Colorado River water is hard and saline. This is due to characteristics of much of the groundwater of the system and probably primarily, due to use of the water for agriculture. Agriculture use produces a great deal of evaporation and hence increased salt content of the natural water, which because of the substrates it flows through, is hard and alkaline, and because of the groundwater, is saline. In the early days, the river was divided into upper and lower basins, and because of the scarcity of water, political considerations have dominated conditions in the Colorado River.

In the book I describe the alterations made to the Colorado River and its tributaries that have affected the aquatic life, particularly in areas where the functional ecosystem is described. The Colorado River has been altered so much that natural ecosystems are extremely rare. The area that seems most natural and for which we have the greatest amount of information is in the reach between Glen Canyon Dam and Hoover Dam above Lake Mead: that is, from Glen Canyon Dam to the beginning of Lake Mead.

The data used in describing the general characteristics of the river as well as of biological ecosystems were obtained largely from the literature. However, with regard to certain species, the data are from collections of the Academy of Natural Sciences of Philadelphia.

Functional ecosystems have been described for some of the tributaries of the Colorado, particularly the ecosystem existing at the mouth of Vasey's Paradise. Ecosystems have also been described for the reaches between Glen Canyon Dam and Lake Mead in the Colorado River; for the Gunnison River; and for the Aravaipa and Sycamore Creeks, tributaries to the Gila River.

Unfortunately, I was not able to obtain sufficient literature to describe the functional ecosystem on the Green River or other areas of the Colorado River system that are in fairly natural conditions. A great variety of literature has been read to bring together the information included in this book. One of the most useful books concerning insects has been *An introduction to the aquatic insects of North America* by R. W. Merritt and K. W. Cummins (Dubuque: Kendall/Hunt Publ. Co., 1996).

In describing the biology of certain sections of the river, I have tried to include the conditions of biology of the riparian areas since they greatly influence the character of the river. Most of the aquatic life in the main stem of the river has been found in the mouths of the tributaries and in selected other areas where conditions were suitable for collection. This is particularly true in the reach between Glen Canyon Dam and Hoover Dam.

The nomenclature used in referring to the various species included in this book is that of the original literature. No attempt has been made to include the latest classification of these various groups of species. I want to thank particularly Dr. Dean W. Blinn and his associates at the University of Arizona for reading the manuscript and providing very valuable sources of information which I have used extensively in creating this volume.

The second part of the book includes descriptions of the salinity problem in the riverine systems, various attempts to control salination, and the political atmosphere that surrounded management of the river.

Also discussed are the problems of quantity and quality of the riverine water, especially salinity. This has been a subject not only of consideration by states through which the rivers flow, but also of the U.S. Congress and of presidents from Franklin D. Roosevelt to Bill Clinton, and continues to be a subject of controversy in Congress.

This book, as well as others in the series, should be very useful to students of rivers and to people of all ages who are interested in conservation and preserving the naturalness of streams. By understanding these systems one can better understand the reasons that it is so important to maintain the naturalness of streams both in structure and in the chemical characteristics of the water. I believe that this book will be useful not only to students but to industrialists and conservationists interested in learning about how various manipulations affect the fauna and flora of rivers. It should also be useful to people interested in the functioning of natural ecosystems in a river.

The book should be of interest to developers and planners, as it will help them to realize the kinds of problems that arise in the development of our Western rivers, where alkalinity and salinity are such a problem. Undoubtedly, the newer methods of drip irrigation and other means of water conservation will do a great deal to extend the water supply of the Colorado River system. However, the development of this section of the country given the scarcity of water in the region poses one of the greatest problems in the United States today.

Acknowledgments

I wish to acknowledge the many people who have made this book possible, particularly those who have done the research that has contributed to this work. Dr. Dean W. Blinn, Regents Professor of Biology at Northern Arizona University, has been of great help in suggesting literature that should be included and also in reading the manuscript. Associated with him is his research associate, Joseph P. Shannon, who assisted in reviewing this volume.

In bringing together the scientific information that formed the basis of this book, I am indebted to my assistant, Rita Kolchinsky. I want to particularly acknowledge Susan Durdu, who has been very helpful as an editor in putting the book together and in doing much of the manual work to make it a success. I am also grateful to the artist, Su Yong, who helped to construct some of the maps and figures. I am particularly grateful to the library of the Academy of Natural Sciences of Philadelphia, particularly Mr. Elliott, the Librarian, who has been very helpful in bringing together the literature necessary for these studies. Besides my colleagues at the Academy, I am very grateful to the personnel of John Wiley & Sons, Inc., particularly Dr. Philip C. Manor, Editor.

This book would not have been possible without the financial help of many people. I want particularly to acknowledge the John and Alice Tyler Award, because the money I received associated with this award has paid for many of the expenses created in the writing of the book. Others who have been of assistance in providing the necessary finances to prepare this large and comprehensive volume include E.I. DuPont Company, the Phoebe W. Haas Charitable Trust, the Procter and Gamble Company, Marion B. Stroud, The Boquet Foundation, and Lewis H. Van Dusen.

I also am deeply appreciative of the time and thoughtful advice given by many of my colleagues and friends in putting together diverse information regarding the chemical, physical, and biological characteristics of the Colorado River system, also in helping me with the political laws and regulations and general discussion that exists about any river, which has such different uses.

CHAPTER 1

General Description
of the Colorado River

INTRODUCTION

The Colorado River, which is 1,397.25 miles in length (2250 km), begins its long journey to the Gulf of California in Grand County, Colorado on the Continental Divide (Ghassemi et al., 1995). It is formed by many small streams that receive an abundance of water from melting snows high in the Rocky Mountains. These snows normally keep the river so full that a portion is siphoned off through the Continental Divide via Adam's Tunnel to the Big Thompson River. This river is the source of irrigation of the South Platte River valley in northeastern Colorado.

At the end of the canyon the swollen Colorado receives Fremont and Escalante drainage from the Great Arid Kaparowitz Plateau. The Colorado River is joined by the Gunnison, which is formed by the North and South Branches. The Uncompahgre River joins the Gunnison, which then joins the Colorado. Farther downstream the San Miguel and Dolores Rivers join the Colorado River in northern Utah.

The Green River, which is the longest tributary of the Colorado River, joins it in Utah. It receives many tributary streams (Figure 1.1). The Green River rises in the Wind River Mountains of Wyoming and is fed by a dozen creeks. From Wyoming the Green flows across the northeastern corner of Utah, receiving still more melted snow from the Uintah Mountains, the only major east–west range in the Rocky Mountain System and the highest point in Utah. Looping through northwestern Colorado, the Green is joined by the Little Snake from the Medicine Bowl Range of Wyoming and the Yampa and the White Rivers of Colorado. Curving back into Utah, the Green gathers from the west the Duchesne, Price, San Rafael, and the Dirty Devil

Figure 1.1. Map of Colorado River basin.

2

Rivers of Utah. Moab Canyon of Utah intersects with the Green River. The next river to join the Colorado is the Escalante River in the mouth of Lake Powell. Lake Powell is a large lake formed behind the Glen Canyon Dam.

The San Juan joins the Colorado in the area of Lake Powell and forms a long estuary in Lake Powell. Many tributary streams join the San Juan in Colorado, New Mexico, and Utah (Figure 1.1). The San Juan is not only Colorado's largest river, but the largest river in New Mexico. Its annual discharge is 2,500,000 acre-ft, which is over twice that of the noted Rio Grande. The headwaters of the San Juan are in southwestern Colorado near Wolf Creek Pass. Flowing south and west it bends through New Mexico for 100 miles (62 km), irrigating a mere 50,000 acres out of 600,000. Rivers draining into the San Juan are the Navajo River, Piedra River, Plata River, Los Piños River, Florida River, Animas River, Mancos River, Aztec Wash, McElmo Creek, and Yellowjacket Canyon, all from Colorado. From New Mexico, the rivers draining into the San Juan are the Chaco, Blanco Canyon, Gallegos Canyon, Jaralosa, Carrico, and Largo Creeks, and from Utah the Montezuma and Gypsum Creeks. The San Juan cuts its way through Utah to join the Colorado. In Arizona, approximately 200 river miles (321.87 km) south of the junction of the Colorado and the San Juan, the Colorado receives the Paria River from Bryce Canyon, Utah.

Vasey's Paradise is a spring in Marble Canyon located between river miles 31 and 32.[1] It drains the eastern part of the Kaibab Plateau. Water cascades down redwall cliffs to terraced slopes at the base of the falls. The abundant vegetation near the spring consists of dense stands of monkeyflower and poison ivy (*Rhus radicans*). Water temperature near its confluence with Colorado River varies from 9 to 18.5°C. The specific conductance ranges from 248 to 420 μS cm^{-1} and the pH is 8.4.

Buck Farm Canyon is an intermittent stream that enters the Colorado River at river mile 41. Little Nankoweap Creek is a small, intermittent stream that enters the Colorado River between river miles 51 and 52. The Little Colorado, which drains 141,155 km^2 in eastern and northern Arizona, enters the Colorado River between river miles 61 and 62. Salt cedar and coyote willow occur near the confluence, while farther upstream, stands of salt cedar and some sparse stands of the common reed (*Phragmites australis*) are found along the margins of the river. The water temperature at the confluence of the Little Colorado was 8.5 to 14.5°C during the period studied. Where the Little Colorado enters the Colorado River, the pH is 7.8 to 7.9. Unkar Creek is a small intermittent stream that enters the Colorado River between river miles 72 and 73. Near the mouth of the creek the vegetation consists of cane beard grass (*Bothriochola barbinodis*), sawgrass (*Cladium californicum*), and the common reed. Clear Creek enters the Colorado River from the north near river mile 84. The specific conductance at the mouth of the creek was 350 to 1700 μS cm^{-1}, and the pH ranged from 8.1 to 8.6. Bright Angel Creek enters the Colorado River on the north rim between river miles 87 and 88. The specific conductance at Bright Angel Creek ranges from 220 to 450 μS cm^{-1} at the mouth. A pH of 8.0 to 8.2 is quite stable at the

---

[1] There is some discussion about the exact location, but the mileage we have used comes from Stevens (1983).

confluence of the creek with the Colorado River. Pipe Creek enters the Colorado River from the south between river miles 88 and 89. The vegetation along the river is sparse and consists of seep willow, Bermuda grass (*Cynodon dactylon*), and scratch grass. Specific conductance ranges from 430 to 590 μS cm^{-1} at the confluence and from 480 to 600 μS cm^{-1} at the upstream site; pH is 8.0 to 8.7 at the mouth and 8.0 to 8.6 upstream. Hermit Creek enters the Colorado River from the south at river mile 95. Its source is a number of springs along the south rim.

Crystal Creek enters the river from the north rim between river miles 98 and 99. No vegetation grew at the confluence of this stream with the Colorado River. Upstream, the canyon is wide, with an open overstory of salt cedar and seep willow. The specific conductance at the mouth was 140 to 474 μS cm^{-1}. Upstream, this value ranged from 240 to 420 μS cm^{-1}. The pH was 8.0 to 8.1. Shinumo Creek enters the Colorado River from the northwest between river miles 108 and 109. White Creek joins Shinumo Creek 5 miles (8.05 km) from the mouth of Shinumo Creek. Occurring along the margins of the stream are seep willow and satintail (*Imperata brevifolia*). The water temperature of Shinumo Creek at its mouth was 8.0 to 20°C. Near the base of the waterfall the temperature varied 10.5°C. Elves Chasm is located between river miles 116 and 117, where Royal Arch Creek enters the Colorado River. Very little vegetation is present at the mouth of the creek.

A short distance upstream, just above the high-water mark of the Colorado River, are several small pools. Vegetation along the margins of these pools consists of seep willow, salt cedar, and redbud. The temperature fluctuates over the seasons and varies from 8.5 to 17.5°C. Specific conductance varies from 429 to 1200 μS cm^{-1} at the mouth and from 525 to 940 μS cm^{-1} upstream. The pH is 8.1 at the mouth.

Stone Creek enters the Colorado River from the north rim between river miles 131 and 132. It is an intermittent stream. Tapeats Creek enters the Colorado River near the north rim at river miles 133 and 134. Tapeats Creek contributes the largest discharge into the Colorado River from the north side of the Grand Canyon. Along the stream margins are the scouring rush (*Equisetum hiemale*), jointed rush (*Juncus articulatis*), cardinal monkeyflower (*Mimulus cardinalis*), and water speedwell (*Veronica aquatica*). Deer Creek enters the Colorado River from the north bank between river miles 136 and 137. Its source is a number of springs. Near the mouth of Deer Creek below the falls, vegetation consists primarily of seep willow, salt cedar, coyote willow, and scratch grass (*Muhlenbergia asperifolia*). The temperature variance is from 12 to 19°C. Specific conductance ranges from 350 to 442 μS cm^{-1} at the mouth. The pH is about 8.3 at the confluence during August and October.

Kanab Creek arises in Cane County, Utah and flows southward approximately 105 km into Arizona, where it enters the Colorado River between river miles 143 and 144. Marginal vegetation at the confluence with the Colorado River consists of a few salt cedars and seep willows. Upstream areas are bordered by salt cedar, bullrush, Arizona grape (*Vitis arizonica*), and cattails (*Typha* spp.). Kanab Creek is a sulfate stream and is low in nitrogen and phosphorus and relatively high in silica. 150 Mile Canyon, a small stream, flows intermittently into the Grand Canyon area between river miles 149 and 150. It is less than 2 miles (3.22 km) in length. Near the mouth of the creek are small pools and they are surrounded by seep willow, saw grass, bull-rush (*Fraxinus* sp.), and scratch grass. Bullrushes, scratch grass, blue-eyed grass (*Sisyrinchium demissum*), and maidenhair fern line the canyon walls above the falls of this creek.

Havasu Creek enters the Colorado River from the south side of the canyon between river miles 156 and 157. Havasu Creek is the only major tributary classified as an impure dolomitic stream. The creek carries carbonate waters dominated by calcium and magnesium and has a silicate concentration second only to that of Diamond Creek. Both nitrogen and phosphorus values are low in Havasu Creek. The temperature at the confluence ranges from 14° to 23.5°C, specific conductance from 660 to 740 μS cm^{-1}, and pH from 8.3 to 9.0.

National Canyon enters the Colorado River from the southeast between river miles 166 and 167. It is a small, intermittent stream. Lava Falls Spring is located between river miles 179 and 180 on the south side of Lava Falls Rapids. Three Springs Canyon enters the Colorado River between river miles 215 and 216. It is an intermittent creek and is relatively short. Vegetation is sparse at the confluence. Vegetation is more apparent upstream. Seep willow, salt cedar, monkeyflower, cattail, and common reed constitute most of the upstream marginal vegetation.

217 Mile Canyon enters through the canyon between river miles 217 and 218 and is a small, intermittent stream. Vegetation is limited to a few seep willows and alkali golden bush.

Diamond Creek enters the Colorado River from the north between river miles 225 and 226. Below the springs Diamond Creek is intermittent. At its mouth it is wide, with little vegetation. Seep willow, cattails, and rabbit foot grass (*Polypogon monspeliensis*) occur at or near the margins of the stream. Water temperature at the mouth ranged from 12 to 27°C. Travertine Falls is located between river miles 230 and 231 on the south side of the canyon. It is a small waterfall approximately 24 m in height and about 200 m from the Colorado River. It forms an intermittent stream. Bridge Canyon is located between river miles 235 and 236. It is a small intermittent creek. Its average gradient is 152 m km^{-1}. No vegetation was observed growing at the edges of the pools.

Above the Hoover Dam at Grand Wash (river mile 276), the river turns sharply right and flows south again, forming the boundary between Arizona and Nevada. In this area it is joined by the Virgin River from the Zion Canyon. The Colorado receives the Bill Williams River, which drains 5400 square miles in Arizona.

Near Yuma, the Colorado receives the Gila River, which rises in the Elk Mountains. It is 630 miles (1010 km) long and one of the most important rivers in America's Southwest. The Gila flows through some of the most arid and desert country in the United States. The Coolidge Dam was constructed in 1928 on the Gila River with a reservoir to hold 6000 million feet (171,428,570 m^3) of water. Arizona's largest river, the Gila drains an area of approximately 56,000 square miles (34,720 km^2), part of which is in New Mexico, where the headwaters of the river are in the Mogollon Mountains.

The Gila River on its course through Arizona receives the San Francisco River, which comes down from the northeast. The Gila alternates its course from southwest to northwest in reaches of 20 to 60 miles (32 to 96 km) through several towns before entering the Gila Indian Reservoir in Pinal County. The chief tributaries of the Gila are the Salt, the San Simon, the San Pedro, the San Carlos, the Santa Cruz, the San Francisco, the Agua Fria, and the Hassayampa Rivers. The Gila also receives numerous intermittent streams and creeks.

The Salt River is a tributary of the Gila formed by the union of the Black and White Rivers in the Mogollon and Black Mountains. Later it is fed by the Tonto Creek from the High Tonto Basin. On the Salt River is the Roosevelt Dam and Reservoir, which irrigates a large portion of southern Arizona.

Downstream from the Hoover Dam is Davis Dam, which has formed Lake Mohave and provides considerable water for irrigation and power generation. The Hoover Dam, on the Colorado River, is mainly for power generation, but also performs other functions.

The Colorado River then flows generally in the southerly direction, forming the borders between California and Arizona. The next humanmade impediment is Parker Dam at the southeastern end of Lake Havasu. The Parker Dam is the beginning of the Colorado Aqueduct, which carries water 389 km across the Mojave Desert and mountains to the city of Los Angeles and other areas of southern California. The two other large reservoirs on the Colorado River are the Imperial and the Laguna Reservoirs. These two reservoirs furnish a large amount of water for irrigation in the desert.

For 80 miles (49.6 km) the Colorado flows through Mexico, separating areas known as Sonora and the mainland from the upper district of the penninsula of lower California. Winding through a maze of salty marshes, deserts, geisers, and dry lakes, it splits into a main channel and tributaries. The river moves slowly into the Gulf of California and sinks its red waters into the blue waters of the ocean.

REGIONS OF THE COLORADO RIVER

The flow of the Colorado River is highly variable. Historical records indicate that it has varied from less than 6 million to more than 20 million acre-feet per year. To meet downstream demands during dry periods, a system of reservoirs has been created which allows sufficient storage water to maintain the flows of the river during these periods. Besides the major reservoirs, there are numerous small reservoirs on the tributaries of the Colorado River (Figure 1.1). These reservoirs now have a combined storage capacity equal to approximately four times the total average annual virgin flow of the Colorado River.

The Colorado Basin has three distinct regions covering 244,000 m^2 or 156,160,000 acres (Brown, 1927).

Upper Basin. The Upper Basin lies in Colorado, New Mexico, Utah, and Wyoming between the Wasatch and Rocky Mountains. This region receives heavy precipitation in the form of snow and provides 75% of the river's flow. In this area the rainfall is less than 10 in. (25.4 cm)/yr over the valleys of the Upper Basin, but irrigation in the valleys has been adapted to agriculture and manufacturing (Brown, 1927).

Plateau Section. The plateau section from Lees Ferry to Grandwash Cliffs has an elevation of 2000 to 8000 ft (609.6 to 2,438.4 m) and is cut by deep gorges and canyons (Brown, 1927). This plateau section contains the area known as the Grand Canyon.

Lower Basin. The Lower Basin lies in Arizona, California, Nevada, and Mexico only a few feet above sea level. It has a precipitation rate of 1.5 to 8 in./yr, with temperatures ranging from 32 to 120°F. A temperature of 130°F has been recorded in the Salton Basin. This basin is separated from the Colorado by a deltic ridge and by the river itself. The basin is below sea level and 100 ft below the Colorado River (Brown, 1927).

The dominant characteristics of the Colorado are its profile, which through much of its course has little slope and is slow moving, and its large silt load. In its 1397-mile (2248.25 km) length, the river falls 8000 ft (2438.4 m). If one eliminates the steep slopes of the headwaters of the Green River, there is a drop of 4000 ft. For half this distance the river flows in a canyon. The Yuma River carries approximately 100,000 acre-ft of silt annually into the Colorado.

DAMS

A large number of dams have been built on the Colorado River system (Figure 1.1). Many of these are on tributaries and are relatively small. These dams were built to supply water used to irrigate farmland. Other dams are placed in deep gorges, where the fall of the water is great. These are rather expensive dams to build but pay for themselves because of the electric power they generate. One of the largest dams is the Hoover Dam, which was authorized in 1928 and finished in 1930. The Cooley Dam, on a tributary of the Gila River, was authorized in 1931 and construction was started in 1933. It was finally finished in the late 1940s. The Flaming Gorge Dam was built on the Green River and the Black Canyon Dam on the Gunnison River. The Red Canyon Dam was built near the Hoover Dam. The Dolores Dam was built on the Animas River, the Yampa Dam on the San Miguel River. As stated above, these dams were constructed primarily for agricultural use or for the generation of power and the supplying of water. The larger dams were built to supply water to the various Compact states (Committee to Review the Glen Canyon Environmental Studies, 1991).

On the main stem of the Colorado River is the Glen Canyon Dam, which is at the head of the reach of the river commonly known as the Grand Canyon. The large lake that it forms is Lake Powell. This lake is approximately 80 miles (128.75 km) long. The Green River enters the Colorado River at the head of Cataract Canyon. The land along the upper Green River is heavily irrigated, as is the land beside the two major tributaries, the Yampa and White Rivers. Other tributaries entering Lake Powell below the Glen Canyon Dam are the Little Colorado River and Kanab Creek.

Lake Mead is a large lake formed by the Hoover Dam. Lake Mojave and Lake Havasu are smaller lakes along the Colorado River.

Besides the Glen Canyon Dam and the Hoover Dam, other dams on the lower Colorado River are the Headgate Rock Dam, the Palo Verde Diversion Dam, the Imperial Dam, and the Morelos Dam in Mexico.

GEOLOGY OF THE WATERSHED

The geology of the Colorado River basin is highly varied. It is composed of igneous, metamorphic, and sedimentary rock types. They range in age from 500 million years to recent alluvial deposits. Structurally, the geology adds anticlines, domes, and faults to the topographic relief and the geohydrology of the region. Sedimentary formations found in the basin were deposited in marine or brackish-water environments. Bedded and disseminated sodium chlorides (halite) and calcium sulfate (gypsum) found as clays have high contents of exchangeable sodium and magnesium (U.S. Department of the Interior, 1997).

Soils

As one might expect, the soils of the basin resemble the geologic formations from which they were derived. The residual soils were derived from shale or sandstone and are generally shallow. These soils contain appreciable soluble mineral content, due to residual and secondary mineral formation from parent material. Upon weathering or irrigation, salts may accumulate on or near the surface, due to evaporation or consumptive use by plants (U.S. Department of the Interior, 1997).

Geology

Sedimentary formations that were deposited in marine or brackish-water environments are found in the basin. Bedded and disseminated sodium chloride (halite) and calcium sulfate (gypsum) are found as clays with high contents of exchangeable sodium and magnesium. The Grand Canyon's older Precambrian rocks are made up of the Vishnu Group, the Zoroaster Plutonic Complex, and the Trinity and Elves Chasm Gneisses. The Vishnu Group is composed of highly metamorphosed schists and gneisses that were originally sediments and minor mafic igneous rocks. These rocks were completely deformed by metamorphism. In the Lower Basin they are imprinted with a well-developed subvertical schistosity (Brown and Babcock, 1974).

Protruding on the Vishnu Group probably during and after regional metamorphism were granite plutons of the Zoroaster Complex. The last episode of activity, probably a hydrothermal rather than a magnetic event, produced pegmatite and aplite dikes and sills. The Trinity and Elves Chasm Gneisses contain layers of the Vishnu formation that grade at their boundaries into the Vishnu Group. These formations demonstrate a metasedimentary origin for these rocks. However, over large areas the gneisses are homogeneous and quartz–monzonite to quartz–dorite in composition. The origin of these formations is unknown; however, their structure gives them an affinity to igneous rocks (Brown and Babcock, 1974).

Zones of intense shearing developed after metamorphism and plutonism. Movement offset metamorphic grades as much as several kilometers vertically in some zones and also caused local retrogressive metamorphism (Brown and Babcock, 1974).

In the reach of the river through Marble and Grand Canyons, the Colorado River cuts along a number of distinct geologic strata, ranging in age from 180 million years (Jurassic) to 1.5 billion years (Precambrian). In Marble Canyon, from Lees Ferry to approximately river mile 51, the Colorado River cuts through late Paleozoic and Metazoic formations, the majority of which are highly resistant to erosion.[2] In this section the steep walls of redwall limestone form a narrow canyon. Between approximately river miles 51 and 75, the canyon widens so that the river cuts through less resistant early Paleozoic and late Permian strata. Generally, this area is composed of mauve limestone and bright angel shale.

Between approximately river miles 75 and 240, the canyon is separated into five natural subdivisions: an upper, a middle, and a lower granite gorge, and two intervening sections where early Paleozoic rock forms vertical cliffs at river level. The upper gorge is approximately 45 km long and extends from river miles 75 to 108. This area is characterized by steep walls of highly resistant Precambrian granites and schists and a narrow (80 m), swiftly moving channel. The Precambrian rock disappears near river mile 116. The middle granite gorge begins at approximately river mile 125. Its walls are less prominent than those of the upper gorge and have only a limited thickness exposed at river level. The canyon again narrows near river mile 140, due to the Precambrian rock at river level being replaced by early Paleozoic strata. The river in this area is similar to Marble Canyon. The lower granite gorge begins at approximately river mile 215 and extends downstream to river mile 240. A relatively large portion of the lower gorge's Precambrian rock is exposed at river level. It forms a V-shaped canyon similar to that in the upper gorge (Hamblin and Rigby, 1968).

BIOLOGY OF THE WATERSHED

The flora and fauna of the Colorado River in the riparian areas is best known in that reach, commonly known as the Grand Canyon, that lies between Glen Canyon Dam and Hoover Dam. The flooding of the riparian area is much less severe than before the dams were built. The woody species of the riparian area are the component vegetation of the landscape and contribute greatly to the support of insects and invertebrates because these areas are less subject to violent floods and droughts; that is, they are more stabilized than before the dams were built (Committee to Review the Glen Canyon Environmental Studies, 1990).

Plants
Native species are a minor component of the overall flora. There are many introduced species in this area. There is an increasing predominance of salt cedar, an increasing abundance of camel thorn (*Alhagicamelorum*), and herbaceous species such as yellow sweet clover (*Melilotus officinalis*) and spiny sow thistle (*Sonchus asper*). Exotics are common along the river. In undisturbed areas they compete poorly with native plants but invade rapidly and increase aggressively when habitat disturbance removes or weakens the native vegetation. Such species are the tamarisk (commonly known as

---

[2] There is some discussion about the exact location, but the mileage we have used comes from Stevens (1983).

salt cedar). In much of the Southwest, salt cedar has out competed native riparian species, especially in disturbed habitats (i.e., those associated with dammed streams). However, it does not seem to have caused the loss of plant species in the Grand Canyon. Salt cedar is usually considered a poor habitat for most wildlife except mourning and white-wing doves (*Zenaida macroura* and *Z. asiatica*). As a species, it attracts insects as well as riparian birds. It is important as a shelter for recreational uses (Committee to Review the Glen Canyon Environmental Studies, 1991).

Another plant of import is camel thorn, a spiny leguminous half-shrub that has invaded several beaches, especially in the upper portions of the canyon, making them virtually unusable for camping. The Russian olive, a small tree spreading into the canyon from other sources, originated from oriental planting. This species has been planted to enhance wildlife. Mourning dove densities in Russian olive habitats were the highest reported for the northern Rio Grande Valley in New Mexico and Texas. Another woody species native to the area is the western honey mosquite, a small tree or subtree occurring in alluvial deposits. This species is important to several riparian birds. Catclaw acacia is a subtree codominant with mosquite; the coyote or sandbar willow is codominant with salt cedar. Next to the Goodding willow, brevy willow is the species preferred by beaver. The Goodding willow, which occurs only at scattered localities along the river, along with Fremont cottonwood, is at least partially prevented from spreading by beavers. Arrow weed is another clonally spreading medium-sized shrub. Seep willow, a large shrub, has two species: *Baccharis saliscifolia* and *B. emoryii*. *Baccharis saliscifolia* is more common along the river and *B. emoryii* along sidestreams. True willow, desert broom, and other medium-sized *Baccharis* occur in the lower canyon. Waterweed (a *Baccharis*), morphologically and ecologically similar to desert broom, is common. Apache plume is a dominant evergreen shrub. This species is relatively common downstream to about km 93.3. Many of the stands of this species are in poor condition, presumably from the lack of water since construction of the Glen Canyon Dam (Committee to Review the Glen Canyon Environmental Studies, 1991).

The herbaceous riparian species that are common native grasses include foxtail broom and Bermuda grass. Many herbaceous plants have been discovered in this reach of the Colorado River. A checklist in 1932 listed approximately 450 species. A second list included 650 species, and a third checklist, 900 species. Most recently, 1400 species of herbaceous plants have been recorded from this general area. Two species new to science were recorded from this area: *Euphorbia aaron-rossii* and *Flaveria mcdougallii* from Cove Canyon, a tributary of the Colorado (Committee to Review the Glen Canyon Environmental Studies, 1991).

Amphibia

In 1970, numerous species of plants and animals were added to the Grand Canyon checklist. Previously, Tomko (1975) added nine species to Gehlbach's 1966 checklist. Amphibians in general are associated with moist habitats. Found were one species of salamander and five species of toads and frogs, four of which were associated with the Colorado River. Woodhouse's toad (*Bufo woodhouseii*) is restricted to the Colorado corridor, and the leopard frog (*Rana* cf. *pipiens*) is restricted to Cardenas Marsh, a postdam marsh and upper lake (Committee to Review the Glen Canyon Environmental Studies, 1991).

Reptiles

The Grand Canyon rattlesnake or pink rattlesnake (*Curtalus varidus abyssus*) occurs frequently in riparian habitats. Seven species of lizards for the riparian zone were listed by Warren and Schwalbe (1985). The stabilized river flow probably resulted in an increase in reptile populations through greater vegetation biomass, providing, in turn, more cover and an increase in insects available (Committee to Review the Glen Canyon Environmental Studies, 1991).

Birds

In the 1970s, Carothers and Johnson (1975) published a new distribution record for 20 birds in the Grand Canyon region, seven of them additions to the fauna list compiled previously. By 1978, the number of species known from the Grand Canyon region had grown to 284, adding more than 100 additional species to the 180 in the first Grand Canyon checklist of Grater (1937). By 1987 this had increased to 303 avian species (Brown et al., 1987), of which 250 or 83% had been recorded from the Colorado River corridor. Populations of nesting birds in the tamarisk approach the highest numbers in temperate North America. Proliferation of the vegetation in the postdam riparian ecosystem is largely responsible for the increase in number of species of birds, the reason being that the birds selected tamarisk over native species. Why this occurs is not clear. Most of the species recorded are insectivorous, and therefore the relatively high insect populations and low human population probably account for their being present. Of the 78 species present reported by Hoffmeister (1971), 40 occurred in the inner gorge, and most of these occurred in the riparian zone restricted to riparian mammals such as beavers, or some lesser category of usage, such as by desert bighorn sheep (Committee to Review the Glen Canyon Environmental Studies, 1991).

Two endangered bird species of the riparian zone are the bald eagle (*Haliaeetas leucocephalus*) and the peregrine falcon (*Falco peregrinus*). Bald eagles have become regular winter visitors since the mid-1980s in the vicinity of Nankoweap Creek, where they feed on fish. The Grand Canyon supports the highest concentration of breeding peregrines in the lower 48 states. A total of 71 different peregrine breeding areas were documented during partial investigation of the region in 1988–1989. A large proportion of their diet consists of riparian birds caught over the riparian and aquatic zones of the inner canyon. This riparian zone, which is rich in species in the Grand Canyon area, supports the food for fish and birds. The riparian area of the Grand Canyon is a very diversified area, and it has become increasingly moist, contrasting with the increasingly xeric conditions of most riverine ecosystems of the Southwest. Thus "riparian vegetation in the canyon is uniquely valuable since the Colorado River in Grand Canyon is the only major riverine system in the Southwest where there has been appreciable increase rather than decrease in riparian vegetation and associated animal populations" (Committee to Review the Glen Canyon Environmental Studies, 1991).

However, it should be pointed out that that the bald eagle (*Haliaeetus leucocephalus*), does not seem to be correlated with prey abundance, biomass patterns, or habitat conditions associated with eagle foraging. Rather, it seems that human activity may in some cases be responsible for lower eagle abundance in some reaches of the river.

However, it should be pointed out that the dam-related effects of clarity, pattern of flow, and water temperature did not override the geomorphological influences on habitat availability. The recovery of the benthos did not seem to be correlated with the abundance of fish but rather with the geomorphological differences in substrate availability between reaches, mediated by the dam and tributary effects on water clarity and the amount of benthos. The action between flow regulation and geomorphology produced a pattern of circuitous recovery of some physical river ecosystems, usually characterized by distance from the dam. This was not true for the benthos. Improving discharge management for endangered native fish population requires detailed understanding of the existing and potential benthic development and trophic interactions. It has been said that many factors are involved in the recovery of the benthos and hence the fish population (Stevens et al., 1997).

Mammals

The vegetation is very important as food and shelter to mammals. In some situations mammals have a profound influence on the vegetation. For example, herbivorous mammals caused severe damage to vegetation by trampling and reducing populations of some plant species by selective foraging prior to the removal of these feral animals from the canyon. For example, the lack of establishment of Goodding willows and Fremont cottonwoods was probably due to beavers. Beavers also eat the smaller coyote willow and even tamarisk. Foraging of beavers could produce a beaver disclimax of some beaches if the beaver population was large enough. One of the most poorly understood vegetational parameters for most woody riparian species is seedling establishment. For such a large and diverse area, Glen Canyon is notably short on endemic, rare species. Two such species are previously undescribed species of flowering plants, *Flaveria mcdougalii* and *Euphorbia aaron-rossii*.

It is very evident that the riparian areas of the Grand Canyon reach of the Colorado River are very important for the aquatic life of the river, and vice versa. Thus, the high diversity of aquatic life helps to support the high diversity of life in the riparian area, particularly birds and amphibia, and vice versa. The aquatic insects of this area are very important to the fish (Committee to Review the Glen Canyon Environmental Studies, 1991).

RIVERINE STRUCTURE

Areas where armoring has not reached equilibrium are more common in the stretch between the Little Colorado and Lake Mead than between the Glen Canyon Dam and the Little Colorado since the lower river has a much larger sediment input. Between Lees Ferry and Separation Rapids, the Colorado exhibits a typical pool-rapid morphology, with the river dropping approximately 677 m over the 384-km distance.

Rapids are constrictions in the river channel caused by the deposition of debris which are typically areas of accelerated flow and high gradient. Often, the velocities are between 3 and 5 m s^{-1} and the gradient is from 0.5 to 2 m per 100 m in the rapids. Fifty percent of the 677-m elevation change occurs in the 161 rapids (Dolan et al., 1978; Leopold, 1969). Separating the sequences of rapids are areas of greater width, slower velocity, and shallower gradients. The velocities in these areas are from 0.3 to 0.6 m s^{-1} and the gradient is 0.5 m per 100 m (Leopold, 1969).

The discharge of the river varies with width and depth. Under the present flow regime the average width is about 200 m, with a range from 80 to 320 m. The average depth of the river is about 10 to 12 m, although the longitudinal profile is highly irregular. The majority of the deeper pools occur immediately below major rapids, where the water flow emerges as a jet directed toward the riverbed. The scouring of the riverbed forms the pools. The deepest pool measured was 35 m. Usually, the shallow portions of the river occur upstream of the rapids, where the river channel is wider.

At the constriction of the river channel, which occurs in the vicinity of rapids and riffles, the larger portion of the downstream flow is directed against one bank, resulting in a strong upstream surge against the opposite bank. At times these backwaters or eddies may occupy 80% of the total width of the river. Often, sediment deposition and erosion take place in these eddies since they typically have a lower velocity than that of the general river.

CHEMISTRY OF THE WATER

The chemistry of Colorado River water is derived from graphs of the characteristics of the water between river mile 0 and river mile 240 from Lees Ferry (Table 1.1). These results clearly indicate that Colorado River water is of medium hardness, containing many trace metals, which are no doubt derived from the geology of the substrate through which the river passes. Chemical data for the water could not be located that were as thorough as those given in these charts, which we have interpreted from tables of Sommerfeld et al. (1976) (Table 1.2).

During the period when the river was studied by Carothers and Minckley (1981a), the temperature ranged between 6 and 15°C. Usually, temperatures increase downstream during the summer months. During 1975 and 1976 the temperatures varied between 7 and 17.2°C. This was in contrast to predam conditions, when the temperature ranged from 0 to 29.5°C. The specific conductance corrected to 25°C varied from 830 to 16,000 μS cm^{-1} in 1977 and 1978. By contrast, predam conductances varied between 318 and 2430 μS cm^{-1} at Lees Ferry, up to 2900 μS cm^{-1} in stretches down river of the Little Colorado. The pH varied between 7.1 and 7.9 during the study period. The major ions in the main stream were $SO_4 > CO_3 >$ chlorides. There was little change in the amounts of these ions between Lees Ferry and Diamond Creek. The ranking major cations were sodium > calcium > magnesium > potassium (Cole and Kubly, 1976).

TABLE 1.1. *Sites Sampled Along the Colorado River in the Grand Canyon National Park and Vicinity*

Site	River Mile
1. Paria River (Lees Ferry)	0.5
2. Vasey's Paradise	31.9
3. Red Wall Cavern	33.0
4. Showerstall Seep	35.5
5. Nautiloid Seep	36.0
6. Buck Farm	41.0
7. Little Nankoweap Creek	51.5
8. Little Colorado River	61.5
9. Cardenas Creek	71.0
10. Unkar Creek	73.0
11. Clear Creek	84.0
12. Bright Angel Creek	87.9
13. Crystal Creek	98.5
14. Shinumo Creek	108.9
15. Elves Chasm	116.5
16. Fossil Rapids	125.0
17. Stone Creek	131.0
18. Tapeats Creek	133.5
19. Deer Creek	136.3
20. Kanab Creek	143.5
21. Olo Creek	145.5
22. The Ledges	152.0
23. Havasu Creek	156.9
24. Prospect Lake	179.0
25. Pumpkin Springs	212.0
26. Three Springs Canyon	215.5
27. Diamond Creek	226.0
28. Travertine Canyon	229.2

Source: After Stevens (1983).

TABLE 1.2. Chemical Characteristics of Colorado River Water 0 to 240 Miles from Lees Ferry

| | Miles from Lee's Ferry | | | | | |
	0–40	40–80	80–120	120–160	160–200	200–240
Calcium (ppm)						
June	50.4, 48	49.6, 47.1, 50.4	51.2, 50.4, 52, 50.4	51.2, 48, 49.6	48	49.6
August	46.4, 47.1	46.4	57.6			67.2
Magnesium (ppm)						
June	25.4, 26.2, 24.2	25.7, 24.6, 26.5	27.3, 26.2 27.3, 26.2	26.5, 25.8, 26.5	27.6	26.5
August	23.5, 233.8	23.1	26.5			28.8
Potassium (ppm)						
June	3.44, 3.40, 3.40	3.44, 3.76, 3.48	3.68, 3.68, 3.68, 3.56	3.56, 3.56, 3.56	3.68	3.68
August	4.48, 3.36	3.36	4.28			4.92
Sodium (ppm)						
June	68.8, 68.8, 63.2	68, 65.6, 73.6	75.2, 76, 77.6, 73.6	74.4, 78.4, 81.6	81.6	74.4
August	65.6, 64	64.8	81.6			84.8
Cadmium (ppb)						
June	0.31, 0.38, 0.15	0.15, 0.15, 0.15	0.15, 0.15, 0.15, 0.15	0.15, 0.15, 0.38	0.15	0.46
August	0.38, 0.92	0.31	0.46			0.54
Cobalt (ppb)						
June	1.9, 0.77, 2.7	0.4, 2.7, 0.4	1.9, 0.77, 1.5, 1.2	0.77, 3.8, 6.5	0.77	2.7
August	1.2, 0.77	1.2	8.8			1.9
Chromium (ppb)						
June	6.5, 6.9, 1.5	4.6, 8.5, 5.8	5.4, 6.9, 5, 4.6	6.2, 5.4, 5.8	5	4.6
August	4.2, 5	5.8	8			1.2

(continues)

15

TABLE 1.2. *Continued*

	Miles from Lee's Ferry					
	0–40	40–80	80–120	120–160	160–200	200–240
Copper (ppb)						
June	8, 4.2, 5.8	3.5, 4.6, 5	4.6, 4.6, 4.2, 4.2	6.5, 5, 5	6.2	4.2
August	5.4, 5.8	8.8	12.3			19.2
Iron (ppb)						
June	160, 160, 200	240, 240, 240	160, 360	120		200
August	80, 40	80	3800			4720
Manganese (ppb)						
June	4.4, 3.6, 4	5.6, 6, 8.8	15.6, 15.6, 10, 0.8	10.8, 8.4, 10.8	8.4	15.6
August	10.4, 10.4	8.8	192			288
Molybdenum (ppb)						
June	7.7, 8.1	8.8	7.3	20.4		5
August	6.5, 6.9	15.4	12.3			19.2
Lead (ppb)						
June	6.2, 5.4, 5.4	1.9, 4.2, 2.7	2.7, 3.1, 2.3, 3.8	1.9, 5, 1.9	4.6	2.3
August	8.1, 4.6	15.4	12.3			19.2
Zinc (ppb)						
June	127, 77	50	38, 50	31		12
August	31, 65	42	31			31

Source: After Sommerfeld et al. (1976).

SEDIMENT LOAD

The suspended solids carried by the Colorado River vary greatly in different sections of the river. They are also variable with different flows. In 1929, Howard estimated that the amount of suspended material carried by the Colorado River as discharged sediments from 1909–1912 and 1912–1916 was approximately 1 million acre-feet for each of the two periods. It is evident that the amount of suspended material varies greatly in different parts of the river. It also varies as much as 25 to 50% at various times of the year. The average quantity of suspended matter in surface samples was about 60% of the quantity of the bottom samples. The amount of suspended sediments is highly variable. Generally, the suspended matter increases with discharge, and the first part of the increased discharge carries the largest amount of silt.

The amount of suspended material varies in different parts of the river. In the period 1925–1927, the suspended load carried past the Grand Canyon was greater than that carried past Topock, Arizona. Both the Grand Canyon and Topock had greater loads than Yuma in 1926–1928. It should be pointed out that the samples collected at Yuma contained matter brought in by the Gila River. The Bureau of Reclamation's yearly measurements for suspended loads past Yuma for 1912–1921 averaged 178 million tons. The suspended load was not exceptionally high at Yuma for 1925–1928.

It was very difficult to make an estimate of the bedload in these areas. In making the measurements of the movement of solids, a movement in the riverbed was observed. Topock and Yuma had considerably greater movement than the sediments in the Grand Canyon. It was apparent that the quantity of material moved during a change in level of a river is large, but how much material is moved as suspended load and how much is moved as bedload is not known (Howard, 1929).

Estimates of the suspended material range from 88,000 to 250,000 acre-ft/yr. The weight per unit volume differed at different parts of the reservoir because larger, heavier particles will settle out at the head of the reservoir. It was evident that the amount of suspended material at different stations varied widely.

The sediment load has been drastically reduced in the river between the dam and the Little Colorado River. This part of the river is now clear throughout most of the year. The median sediment load at Lees Ferry has been reduced by a factor of about 200, from 1500 ppm to 7 ppm. The reduction in the sediment load is less apparent downstream because of sediment inputs from tributaries and the erosion of predam fluvial terraces. At Phantom's Reach the sediment load had been reduced by only a factor of 3.5. Currently, the Little Colorado River contributes the majority of the sediment to the Colorado River between Lees Ferry and Lake Mead. This marked alteration in patterns of seasonal flow has changed the alluvial morphology of the river. The river has eroded predam terraced deposits at many locations because large gravel and rubble have been exposed that cannot be dislodged. Erosion no longer occurs. This process, referred to as *armoring,* is especially prevalent between the dam and Lees Ferry.

ASSOCIATIONS OF AQUATIC ORGANISMS:
GLEN CANYON DAM TO HOOVER DAM

The associations of aquatic organisms are best known between Glen Canyon Dam and the beginning of Lake Mead. Suspended solids carried by the flow of the river vary greatly between winter and summer. During torrential winter floods of tributaries, it might be greater than 2832 m^3 s^{-1}, whereas during the summer months at low flows it was only 113 m^3 s^{-1} of sediments. The average suspended sediment load in the lower Colorado before impoundment was approximately 3.5 times greater than after construction of the Glen Canyon Dam.

Algae

The algae found in the Grand Canyon reach of the river were either attached or free-floating. Many of the free-floating forms seemed to be part of attached vegetation that had been broken off by the flow. The areas of algae attachment were some of the scoured rock faces in areas of rapids and cataracts and fine sediments in backwater, usually along the inner side of river bends. Various submerged macrophytes served as the surface for attachment of many algae. Dominant diatoms were *Diatoma vulgare* (Bory), *Gomphonema olivaceum* (Lyngb Kütz), *Navicula viridula* (Kütz), *Synedra ulna*, *Nitzschia* (Ehr), and two species of *Surirella*. They were collected in the Colorado River in the stretch between the two dams (Blinn and Cole, 1991) (Table 1.3).

Crayton and Sommerfeld (1978) reported 127 species of phytoplankton in the Colorado River. Many of these they believed to be detached species dislodged during widely fluctuating river levels. Phytoplankton (about 58%) were composed of diatoms, of which the dominant species were *Diatoma vulgare*, *Rhoicosphenia curvata* (Kütz) Grun, and *Cocconeis pediculus* (Ehr). Many of these originated from Lake Powell. For example, of the 20 dominant algal species listed for Warm Creek Bay in Lake Powell, eight species were found to be common to both the river and the lake. The cell densities of diatom species in the Colorado River after impoundment were much lower than the cell densities prior to the impoundment of the river (Blinn and Cole, 1991).

Of the algae, *Cladophora glomerata* (L.) Kütz was the dominant attached filamentous green algae in the canyon, especially between Glen Canyon Dam and the Paria River and at the mouths of tributaries. The free-flowing Colorado River during high flows also carried a lot of sestonic stream drift, including algae. Usher and Blinn (1990) estimated that *Cladophora glomerata* was the dominant alga for sites at and above Lees Ferry: 144 g m^{-2}, compared with 17.2 g m^{-2} at the mouth of various tributaries below Lees Ferry. The relatively high amount of biomass of *Cladophora glomerata* in the upstream tailwater sites may be a result of stable rock faces for attachment and nutrient-enriched waters. At Lees Ferry there were often abrupt drops in population size of *Cladophora glomerata*, which were probably due to episodes of desiccation and reduced input of nutrients. *Cladophora* thrives under continuously submerged clear-water habitats (Shaver et al., 1997). *Cladophora* growths present on the downstream side of rocks were often absent on the upstream side of

(text continues on page 37)

TABLE 1.3. *Species List: Grand Canyon National Park and Vicinity*

Taxon[a]	V	M,S	G,R	Source[c]	Feeding Habits[d]
		Substrate[b]			
SUPERKINGDOM PROKARYOTAE					
KINGDOM MONERA					
Division Cyanophycota					
Class Cyanophyceae					
Order Chamaesiphonales					
Family Chamaesiphonaceae					
Chamaesiphon incrustans, 1–27				CS	
Chamaesiphon sp.				CBT	
Order Chroococales					
Family Chroococaceae					
Aphanocapsa musicola		×		CBT	
Aphanocapsa sp.		×		CBT	
Chroococcus minor				CBT	
C. minutus				CBT	
C. turgidus				CBT	
Chroococcus sp., 1–27				CS	
Gleocapsa polydermatica				CBT	
Gleothece sp.		×		CBT	
Merismopedia glauca			×	CBT	
M. punctata			×	CBT	
Order Nostocales					
Family Microchaetaceae					
Microchaete elongata				CBT	
Family Nostocaceae					
Anabaena oscillariodes		×		CBT	
Anabaena spp.		×		CBT	
Nodularia spumigena, 1–27				CS	
Nostoc hatei	×	×		CBT	
N. paludosum	×	×		CBT	
N. punctiforme	×	×		CBT	
N. verrucosum	×	×		CBT	
Nostoc spp.	×	×		CBT	
Family Oscillatoriaceae					
Gleotrichia intermedia	×			CBT	
Katagnymene pelagica				CBT	
(= *Microcoleus lyngbyaceus*)					
Lyngbya aerugineo-caerulea			×	CBT	
L. aestaurii			×	CBT	
L. allegori			×	CBT	
L. cryptovaginata			×	CBT	
L. digueti			×	CBT	
L. epiphytica			×	CBT	
L. hieronymusii			×	CBT	
L. limnetica			×	CBT	
L. major			×	CBT	
L. martensiana			×	CBT	
L. mesotrichia			×	CBT	
L. nordgardhii			×	CBT	
L. perelegans			×	CBT	
L. statina			×	CBT	
L. versicolor			×	CBT	
Lyngbya spp.			×	CBT	
Oscillatoria acuminata		×	×	CBT	

(continues)

TABLE 1.3.   *Continued*

Taxon[a]	V	M,S	G,R	Source[c]	Feeding Habits[d]
		Substrate[b]			
O. agardhii		×	×	CBT	
O. amoena		×	×	CBT	
O. amphibia		×	×	CBT	
O. amphigranulata		×	×	CBT	
O. angustissima		×	×	CBT	
O. articulata, 1–27		×	×	CS	
O. chalybea		×	×	CBT	
O. claricentrosa		×	×	CBT	
O. cortiana		×	×	CBT	
O. foreaiu		×	×	CBT	
O. fremyii		×	×	CBT	
O. hamelii		×	×	CBT	
O. jasorvensis		×	×	CBT	
O. lemmermannii		×	×	CBT	
O. limnetica		×	×	CBT	
O. limosa, 1–27		×	×	CS	
O. migro-viridis		×	×	CBT	
O. mougeotii		×	×	CBT	
O. nigra, 1–27		×	×	CS	
O. obscura		×	×	CBT	
O. okeni		×	×	CBT	
O. proteus		×	×	CBT	
O. pseudogeminata		×	×	CBT	
O. quadripunctulata		×	×	CBT·	
O. rubescens		×	×˙	CBT	
O. schultzii		×	×	CBT	
O. simplicissima		×	×	CBT	
O. splendida		×	×	CBT	
O. subbrevis		×	×	CBT	
O. tanganyikae		×	×	CBT	
O. tenuis		×	×	CBT	
O. tenuis var. *tergestina*		×	×	CBT	
O. trichoides		×	×	CBT	
Oscillatoria spp.		×	×	CBT	
Phormidium ambiguum		×	×	CBT	
P. anomala		×	×	CBT	
P. corium var. *constrictum*		×	×	CBT	
P. dimorphum		×	×	CBT	
P. mucosum		×	×	CBT	
P. retzii		×	×	CBT	
P. tenue		×	×	CBT	
Spirulina labyrinthiformis		×	×	CBT	
S. major, 1–27		×	×	CS	
S. subsalsa, 1–27		×	×	CS	
S. subtilissima		×	×	CBT	
Spirulina sp., 1–27		×	×	CS	
Symploca sp.				CBT	
Family Scytonemataceae					
Scytonema alatum	×	×		CBT	
S. rivulare	×	×		CBT	
Order Stigonemales					
Family Stigonemaceae					
Stigonema hormoides				CBT	

TABLE 1.3. *Continued*

	Substrate[b]				
Taxon[a]	V	M,S	G,R	Source[c]	Feeding Habits[d]
SUPERKINGDOM EUKARYOTAE					
KINGDOM PLANTAE					
Subkingdom Thallobionta					
Division Chlorophycota					
Order Chaetophorales					
Family Chaetophoraceae					
Cloniophora sp., 1–27				CS	
Stigoclonium flagelliferium, 1–27			×	CS	
S. pachydermum			×	CBT	
Stigoclonium sp., 1–27			×	CS	
Tetraedron sp., 1–27			×	CS	
Order Chlorococcales					
Family Chlorococcaceae					
Chlorococcum spp.	×	×		CBT	
Family Hydrodictyaceae					
Pediastrum boryanum	×	×		CBT	
P. duplex var. *clathratum*, 1–27	×	×		CS	
P. integrum	×	×		CBT	
P. integrum var. *scutum*	×	×		CBT	
Family Oocystaceae					
Cerasterias staurastroides, 1–27				CS	
Lagerheimie subsalsa, 1–27				CS	
Oocystis crassa		×		CBT	
O. elliptica		×		CBT	
O. solitaria		×		CBT	
Family Scenescedesmaceae					
Scenedesmus bijuga, 1–27		×		CS	
S. opoliensis, 1–27		×		CS	
S. quadricauda var. *maximus*, 1–27		×		CS	
Order Cylindrocapsales					
Family Cylindrocapsales					
Cylindrocapsa sp.				CBT	
Cylindrotheca gracilis				CBT	
Order Tetrasporales					
Family Tetrasporaceae					
Tetraspora cylindrica	×		×	CBT	
Tetraspora sp.	×		×	CBT	
Order Trentepohliales					
Family Trentepohliaceae					
Gongrosira lacustris				CBT	
Leptosira sp., 1–27				CS	
Trentepohlia aurea				CBT	
Order Ulotrichales					
Family Microsporaceae					
Microspora floccosa, 1–27			×	CS	
M. loefgrenii			×	CBT	
M. pachyderma			×	CBT	
Microspora sp.			×	CBT	
Ulothrix aequalis	×			CBT	
U. cylindricum	×			CBT	
U. subtilissima	×			CBT	
U. tenerrima	×			CBT	
U. tenuissima	×			CBT	

(continues)

TABLE 1.3. *Continued*

Taxon[a]	Substrate[b] V	M,S	G,R	Source[c]	Feeding Habits[d]
U. variabilis	×			CBT	
U. zonata	×			CBT	
Ulothrix sp., 1–27	×			CS	
Family Monostromataceae					
Gomontia sp., 1–27				CS	
Order Oedogoniales					
Family Oedogoniaceae					
Oedogonium spp.	×		×	CBT	
Order Cladophorales					
Family Cladophoraceae					
Cladophora fracta	×	×	×	CBT	
C. glomerata	×	×	×	CBT	
Rhizoclonium hieroglyphicum	×	×	×	CBT	
R. hookeri	×	×	×	CBT	
Order Zygnematales					
Family Desmidiaceae					
Closterium acerosum var. *elongatum*	×		×	CBT	
Closterium spp.	×		×	CBT	
Cosmarium spp.	×	×		CBT	
Staurastrum sp., 1–27		×		CS	
Family Zygnemataceae					
Mougeotia spp.	×		×	CBT	
Mougeotiopsis sp., 1–27	×		×	CS	
Spirogyra spp.		×	×	CBT	
Zygnema spp.			×	CBT	
Class Bacillariophyceae					
Order Eupodiscales					
Family Coscinodiscaceae					
Coscinodiscus denarius, 20			×	CBT	
Cyclotella atomus, 4	×	×	×	CBT	
C. meneghiniana, throughout Colorado River	×	×	×	CBT	
C. michiganiana, Lees Ferry, mile 19	×		×	CBT	
Melosira granulata, 18, 23	×	×	×	CM	
M. varians, widespread	×	×	×	CBT	
Order Biddulphiales					
Family Biddulphiaceae					
Biddulphia laevis, 15, 27	×		×	CBT	
Order Fragilariales					
Family Fragilariaceae					
Asterionella formosa, upper Colorado River			×	CBT	
Diatoma anceps, 1–27			×	CS	
D. elongatum, 1–27			×	CS	
D. hiemale, 1–27			×	CS	
D. hiemale var. *mesodon*, throughout Glen Canyon, especially 18			×	CB	
D. vulgare, widespread			×	CBT	
D. vulgare var. *breve*, 1, 11, 18, 19			×	CBT	
D. vulgare var. *linearis*, 14			×	CBT	
Fragilaria aequalis, 1–27				CS	
F. brevistriata, 1–27				CS	
F. brevistriata var. *inflata*, 9				CBT	
F. capucina, 15, 18, 23	×			CBT	
F. capucina var. *mesolepta*, 15, 23, 27	×			CBT	

TABLE 1.3. *Continued*

Taxon[a]		Substrate[b]		Source[c]	Feeding Habits[d]
	V	M,S	G,R		
F. construens, 1–27		×	×	CS	
F. construens var. *venter*, 27		×	×	CBT	
F. crotonensis, Lake Powell, upper Colorado River			×	CS	
F. intermedia, 1–27			×	CS	
F. leptostauron, 12, 18		×		CBT	
F. leptostauron var. *dubia*, 8		×		CBT	
F. vaucheriae, widespread	×	×	×	CBT	
F. virescens, 1–27		×		CS	
Meridion circulare, 15, 19, 27	×	×		CBT	
Opephora ansata, 25		×	×	CBT	
Synedra acus, common throughout Glen Canyon, abundant 13		×	×	CBT	
S. affinis, 15		×	×	CBT	
S. delicatissima var. *angustissima* upper Colorado River, probably remnant from Lake Powell			×	CBT	
S. goulardii, 18		×	×	CBT	
S. incisa, 18, 27		×	×	CBT	
S. mazamaensis, single collection at 11		×	×	CBT	
S. miniscula var. *longa*		×	×	CBT	
S. nana, 1–27			×	CS	
S. pulchella var. *lacerata*, 20	×	×	×	CBT	
S. rumpens	×		×	CBT	
S. rumpens var. *familiaris*, 21	×		×	CBT	
S. socia, widespread, high densities above little Colorado River		×	×	CBT	
S. tenera var. *genuina*, 1–27	×		×	CS	
S. ulna, widespread	×	×	×	CBT	
S. ulna var. *constracta*, 2	×	×	×	CBT	
Order Achnanthales					
Family Achnanthaceae					
Achnanthes affinis, widespread	×	×	×	CBT	
A. clevei, 1, 9	×	×	×	CBT	
A. coarctata, 6, 14	×	×	×	CBT	
A. deflexa, 15			×	CBT	
A. exigua var. *heterovalva*, widespread, especially 2		×	×	CBT	
A. flexella, Lees Ferry		×	×	CBT	
A. lanceolata, widespread in Glen Canyon	×	×	×		
A. lanceolata var. *dubia*, widespread in Glen Canyon	×	×	×	CBT	
A. lanceolata var. *omissa*, 19	×	×	×	CBT	
A. linearis, N., 1, 8, 12, 14, 18, 19, 20, 23	×	×	×	CBT	
A. linearis forma. *curvata*, widespread in Glen Canyon	×	×	×	CBT	
A. linearis var. *pusilla*, widespread in Glen Canyon	×	×	×	CBT	
A. microcephala, widespread in Glen Canyon		×	×	CBT	
A. minutissima, widespread	×	×	×	CBT	
A. sublaevis var. *crassa*, 1, 18, 19, 23	×	×	×	CBT	
A. wellsiae, mile 48.9	×	×	×	CBT	

(continues)

TABLE 1.3. *Continued*

Taxon[a]	V	M,S	G,R	Source[c]	Feeding Habits[d]
Achnanthidium sp., 1–27				CS	
Cocconeis diminuta, 2, 12, 15, 19			×	CBT	
C. pediculus, widespread	×		×	CBT	
C. placentula, 12, 14, 18, 19, 23	×		×	CB	
C. placentula var. *euglypta*, widespread	×		×	CBT	
C. placentula var. *lineata*, widespread	×		×	CBT	
Rhoicosphenia curvata, widespread	×		×	CBT	
Order Naviculales					
Family Cymbellaceae					
Amphora adnata	×		×	CBT	
A. arizonica	×		×	CBT	
A. coffeiformis, 25	×		×	CBT	
A. ovalis, 6	×		×	CBT	
A. ovalis var. *pediculus*, widespread	×		×	CBT	
A. perpusilla, 8, 12, 18	×		×	CBT	
A. veneta, 27	×		×	CBT	
Cymbella affinis, widespread	×		×	CBT	
C. affinis var. *bipunctata*	×		×	CBT	
C. amphicephala, widespread	×		×	CBT	
C. aspera, 1–27	×		×	CS	
C. caespitosa var. *ovata*, 1–27	×		×	CS	
C. cistula, 15	×		×	CBT	
C. cymbiformis var. *nonpunctata*, 15	×		×	CB	
C. hungarica, 1–27	×		×	CS	
C. incerta, 1–27	×		×	CS	
C. laevis, widespread in small numbers	×		×	CBT	
C. leptoceros, 19	×		×	CBT	
C. magnapunctata	×		×	CBT	
C. mexicana	×	×		CBT	
C. microcephala, 1, 8, 12, 19, 23	×		×	CBT	
C. microcephala var. *crassa*, 1	×	×		CBT	
C. minuta, widespread	×		×	CBT	
C. norvegica, 6, 17	×		×	CBT	
C. parva	×		×	CBT	
C. prostata, widespread	×	×		CBT	
C. pusilla, 13, 15	×		×	CBT	
C. sinuata, 1	×	×		CBT	
C. tumida, widespread		×	×	CBT	
C. tumidula, 15		×	×	CBT	
C. turgida, 1–27			×	CS	
C. ventricosa, 1–27		×	×	CS	
C. ventricosa var. *semicircularis*, 1–27		×	×	CS	
Family Entomoneidaceae					
Entomoneis alata, 15, 27				CBT	
E. palludosa, 15, 27				CBT	
Family Gomphonemaceae					
Gomphoneis herculeana, widespread	×			CBT	
Gomphonema acuminatus	×			CBT	
G. affine, 3, 18, 23	×			CM	
G. affine var. *insigne*, 20	×			CBT	
G. grunowii, 23	×			CBT	
G. intracatum, 1, 15	×			CBT	
G. intracatum var. *vibrio*, 20	×			CBT	

TABLE 1.3. *Continued*

Taxon[a]	V	M,S	G,R	Source[c]	Feeding Habits[d]
G. olivaceum, 20	×			CBT	
G. parvulum, widespread	×			CBT	
G. subclavatum, widespread	×			CBT	
G. truncatum, widespread	×			CBT	
Family Naviculaceae					
Amphipleura pellucida, widespread		×		CBT	
Anomoeoneis exilis, 1–27		×		CS	
A. serians var. *brachysira*, 1–27		×		CS	
A. vitrea, 11, 27		×		CBT	
Caloneis amphisbaena, 23		×		CM	
C. bacillaris var. *thermalis*, 8, 10, 16		×		CBT	
C. bacillum, widespread		×		CBT	
C. backmanii, mile 19		×		CBT	
C. hyalina, 9		×		CBT	
C. pulchra var. *brevistriata*		×		CBT	
(= *C. silicula* var. *brevistriata*), 15					
C. silicula, 20		×		CM	
C. silicula var. *limosa*, 21		×		CM	
C. ventricosa var. *truncatula*, 20, 22		×		CBT	
Diploneis elliptica, 2	×			CBT	
D. oblongella, 2	×			CBT	
D. oculata, 2, 8	×			CBT	
D. puella, widespread	×			CBT	
D. smithii var. *dilatata*, 15	×			CBT	
Frustulia vulgaris, widespread		×	×	CB	
Gyrosigma spencerii, 20		×	×	CBT	
G. spencerii var. *curvula*, 6, 11, 15		×	×	CBT	
Mastagloia elliptica var. *danseii*, 15				CBT	
M. gravillei, 22				CBT	
M. smithii, widespread				CBT	
M. smithii var. *amphicephala*, 6, 11				CBT	
M. smithii var. *lacustris*, widespread				CBT	
Navicula accomoda, 12, 20		×		CBT	
N. angelica, 1–27		×		CS	
N. angelica var. *subsalsa*, widespread		×		CBT	
N. arvenensis, widespread		×		CBT	
N. bacillum, 11				CBT	
N. cari, 27				CB	
N. cincta, 9, 20		×		CBT	
N. cocconeiformis, 12				CBT	
N. confervaceae, 24	×		×	CB	
N. cryptocephala, widespread	×	×	×	CBT	
N. cryptocephala forma. *minuta*, widespread	×	×	×	CBT	
N. cryptocephala var. *veneta*, widespread	×	×	×	CBT	
N. cuspidata, sporadically observed		×	×	CBT	
throughout the Glen Canyon					
N. cuspidata var. *major*, 23		×	×	CBT	
N. decussis, widespread		×	×	CBT	
N. densestriata, 8				CBT	
N. dicephala		×		CBT	
N. eulineata				CBT	
N. exigua, 21		×		CB	
N. globulifera, 26				CB	

(continues)

TABLE 1.3. *Continued*

Taxon[a]	V	M,S	G,R	Source[c]	Feeding Habits[d]
		Substrate[b]			
N. graciliodes, widespread		×		CB	
N. gregaria, 20		×		CBT	
N. grimmei, 20				CBT	
N. halophila, 15				CB	
N. lanceolata, 20, 23		×	×	CBT	
N. laterostrata, 23				CM	
N. longirostris, 9, mile 115				CBT	
N. minima, widespread			×	CBT	
N. miniradiata				CBT	
N. minuscula, 5, 21, 23				CBT	
N. mutica, widespread	×	×	×	CBT	
N. mutica var. *cohnii*, 13, 15, 18, 27	×	×	×	CBT	
N. mutica var. *stigma*, 15	×	×	×	CBT	
N. mutica var. *undulata*, 15	×	×	×	CBT	
N. notha, 3, 27				CBT	
N. pelliculosa, 20			×	CBT	
N. pseudoreinhardtii, 11, mile 134				CBT	
N. pupula, 23, 25, 26, mile 134		×		CBT	
N. pupula var. *capitata*, 1		×		CM	
N. pupula var. *rectangularis*, widespread		×		CBT	
N. radiosa, widespread	×	×		CBT	
N. radiosa var. *tenella*, widespread	×	×		CBT	
N. secreta var. *apiculata*, somewhat widespread, especially 12, 15			×	CBT	
N. seminulum var. *hustedtii*, 24		×		CB	
N. silicificata				CBT	
N. subtilissima, 2, 6		×		CBT	
N. symmetrica, 6, 23		×	×	CB	
N. tridentula, 6, 14				CBT	
N. tripunctata, widespread	×		×	CBT	
N. tripunctata var. *schizonemoides*, 6, 9	×		×	CBT	
N. tuscula, 1				CBT	
N. viridula, 8, 20, 21	×	×	×	CBT	
N. viridula var. *rostellata*, 23, 27	×	×	×	CBT	
N. zanoni, widespread, most common in 2, 14, 18				CBT	
Neidium binode, 15				CBT	
N. dubium, 1–27		×	×	CS	
N. dubium f. *constrictum*, 15, mile 19		×	×	CBT	
Pinnularia appendiculata, 20, 25		×		CBT	
P. borealis var. *rectangularis*, 15		×		CBT	
P. brebissoni, 19		×		CBT	
P. divergentissima, 6		×		CBT	
P. prescottii		×		CBT	
P. substomatophora, 24			×	CB	
Pinnularia sp., 1–27				CS	
Pleurosigma delicatulum, 15, 23				CBT	
Rhopalodia gibba, widespread				CBT	
R. gibba var. *ventricosa*, 15				CBT	
R. gibberula var. *vanheurckii*, widespread				CBT	
Scoliopleura peisonis, 15				CBT	
Stauroneis anceps, 15	×		×	CBT	
S. amphioxys var. *rostrata*	×		×	CBT	

TABLE 1.3. *Continued*

Taxon[a]	Substrate[b]			Source[c]	Feeding Habits[d]
	V	M,S	G,R		
S. smithii, 15	×		×	CBT	
Family Plagiotropidaceae					
Plagiotropis lepidoptera, 27				CBT	
Order Epithemiales					
Family Epithemiaceae					
Denticula elegans, widespread	×		×		CBT
D. rainerensis, 2, 4, others	×		×	CBT	
Epithemia adnata, 2, 6				CBT	
E. argus, 17, 28				CB	
E. argus var. *alpestris*, 2, 6				CBT	
E. argus var. *longicornis*, widespread				CBT	
E. sorex, widespread				CBT	
E. turgida, upper reaches of Colorado River, below Glen Canyon, but not in Lake Powell area				CBT	
Order Bacillariales					
Family Nitzschiaceae					
Bacillaria paradoxa, 27	×	×		CBT	
Hantzschia amphioxys, widespread	×			CBT	
H. amphioxys forma. *capitata*	×			CBT	
Nitzschia accedans, 12, 14, 15, 21				CB	
N. acicularis, widespread		×		CBT	
N. acicularis var. *closterioides*, 23		×		CB	
N. acuta, 14		×	×	CBT	
N. amphibia, widespread			×	CBT	
N. angustata, 8, 23; mile 34.5				CBT	
N. angustata var. *acuta*, 14, 20				CBT	
N. apiculata, widespread	×			CBT	
N. bicrena, 14				CBT	
N. bita, 15				CBT	
N. capitellata, 20		×	×	CBT	
N. communis, widespread				CBT	
N. denticulata, 2				CBT	
N. dissipata, widespread	×	×	×	CBT	
N. filiformis, 12, 21	×	×		CB	
N. fonticola, 20				CBT	
N. frustulum, widespread	×	×	×	CBT	
N. frustulum var. *perpusilla*, 18	×	×	×	CBT	
N. gracilis, 15				CBT	
N. hungarica, 20				CBT	
N. hybrida, 19				CBT	
N. kutzingiana, widespread			×	CBT	
N. lacunarum, 4				CBT	
N. linearis, widespread				CBT	
N. littoralis var. *tergestina*, 20				CBT	
N. microcephala				CBT	
N. palea, 1, 12, 18, 19, 20, 23	×	×	×	CBT	
N. parvula, 23				CB	
N. pseudolinearis				CBT	
N. recta, 23				CBT	
N. romano, 2, 12, 15		×		CM	
N. scalpelliforma				CBT	
N. sigma, 8, 19, 20	×		×	CBT	

(continues)

TABLE 1.3. *Continued*

Taxon[a]	Substrate[b]			Source[c]	Feeding Habits[d]
	V	M,S	G,R		
N. sigmoidea, 15				CB	
N. sinuata var. *tabellaria*, 14				CBT	
N. tryblionella var. *calida*, 20, 23		×		CBT	
N. tryblionella var. *levidensis*, 21, 23		×		CB	
N. vermicularis, 14			×	CBT	
Nitzschia spp., 1, 8, 12, 19, 23					
Order Surirellales					
Family Surirellalaceae					
Campylodiscus noricus var. *hibernica*, 2				CBT	
Cymatopleura solea, 15		×		CBT	
Surirella angustata, widespread		×		CBT	
S. brightwellei, widespread		×		CBT	
S. ovalis, 8				CBT	
S. ovata, widespread, esp. 20		×		CBT	
S. ovata var. *africana*, 14		×		CBT	
S. ovata var. *pinnata*, 20		×		CBT	
S. patella, 22				CBT	
S. striatula, 11, 15, 27		×		CBT	
S. striatula var. *parva*		×		CBT	
Division Rhodophycota					
Class Rhodophyceae					
Order Nemalionales					
Family Batrachospermaceae					
Batrachospermum sp.			×	CBT	
Division Chromophycota					
Class Xanthophyceae					
Tetragoniella sp., 1–27				CS	
Order Vaucheriales					
Family Tribonemataceae					
Tribonema utriculosum, 1–27				CS	
Family Vaucheriaceae					
Vaucheria geminata				CBT	
V. sessilis				CBT	
Vaucheria spp.				CBT	
Class Dinophyceae					
Order Peridinales					
Family Ceratiaceae					
Ceratium carolinianum, 1–27			×	CS	
C. hirudinella, 1–27			×	CS	
Subkingdom Embryobionta					
Division Magnoliophyta					
Class Magnoliopsida					
Subclass Dilleniidae					
Order Malvales					
Family Malvaceae					
Iliamna grandiflora				CM	
Sida nederacea				CM	
Sidalcea neomexicana				CM	
Order Theales					
Family Elatinaceae					
Elatine brachysperma				CM	
E. triandra				CM	

TABLE 1.3. *Continued*

Taxon[a]	Substrate[b]			Source[c]	Feeding Habits[d]
	V	M,S	G,R		
Family Guttiferae					
Hypericum anagalloides				CM	
H. formosum				CM	
Order Violales					
Family Loasaceae					
Eucidne urens				CM	
Family Tamaricaceae					
Tamarix aphylla				CM	
T. pentandra				CM	
Order Primulales					
Family Primulaceae					
Androsace occidentalis				CM	
A. septentrionalis				CM	
Dodecatheon alpinum				CM	
Samolus parviflorus				CM	
Subclass Rosidae					
Order Proteales					
Family Elaeagnaceae					
Elaeagnus angustifolia				CM	
Order Myrtales					
Family Onagraceae					
Epiloium adenocaulon				CM	
E. halleanum				CM	
E. hornemanni				CM	
E. saximontanum				CM	
Oenothera flava				CM	
O. hookeri				CM	
O. longissima				CM	
Order Apiales					
Family Umbrelliferae (= Apiaceae)					
Berula erecta				CM	
Caucalis microcarpa				CM	
Cicuta douglasii				CM	
Conium maculatum				CM	
Perideridia parishii				CM	
Order Cornales					
Family Cornaceae					
Cornus stolonifera				CM	
Subclass Asteridae					
Order Scrophulariales					
Family Oleaceae					
Fraxinus anomala				CM	
F. cuspidata var. *macropetala*				CM	
F. pennsylvanica velutina				CM	
Order Gentianales					
Family Apocynaceae					
Apocynum cannabinum				CM	
A. sibiricum var. *salignum*				CM	
A. suksdorfii				CM	
Family Asclepiadaceae					
Asclepias subverticillata				CM	
Sarcostemma cynanchoides				CM	
Family Gentianaceae					
Centarium calycosum				CM	

(continues)

TABLE 1.3.   *Continued*

Taxon[a]	Substrate[b]			Source[c]	Feeding Habits[d]
	V	M,S	G,R		
Gentiana affinis				CM	
G. parryi				CM	
Order Solanales					
Family Convolvulaceae					
Convolvulus arvensis				CM	
Cuscuta campestris				CM	
C. coryli				CM	
C. indecora				CM	
Family Hydrophyllaceae					
Phacelia magellanica				CM	
Family Polemoniaceae					
Collomia linearis				CM	
Family Solanaceae					
Datura meteloides				CM	
Nicotiana glauca				CM	
N. trigonophylla				CM	
Solanum douglasii				CM	
S. nodiflorum				CM	
Order Lamiales					
Family Boraginaceae					
Hackelia floribunda				CM	
Heliotropium curassavicum				CM	
Lappula redowskii				CM	
Mertensia franciscana				CM	
Family Labiatae (= Lamiales)					
Mentha arvenis var. *villosa*				CM	
M. spicata				CM	
Nepeta cataria				CM	
Prunella vulgaris				CM	
Family Verbenaceae					
Phyla cuneifolia				CM	
Verbena bractea				CM	
V. macdougalii				CM	
Order Scrophulariales					
Family Scrophulariaceae					
Besseya arizonica				CM	
B. plantaginea				CM	
Castilleja confusa				CM	
Limosella acaulis				CM	
L. aquatica				CM	
Maurandya antirrhiniflora				CM	
Mimulus cardinalis				CM	
M. guttatus				CM	
M. nasutus				CM	
M. primuloides				CM	
M. rubellus				CM	
Penstemon rydbergii				CM	
P. virgatus				CM	
Penstemon sp.				CM	
Veronica americana				CM	
V. anagallis-aquatica				CM	
V. serphyllifolia var. *borealis*				CM	

TABLE 1.3. *Continued*

Taxon[a]	Substrate[b]			Source[c]	Feeding Habits[d]
	V	M,S	G,R		
Order Plantaginales					
Family Plantaginaceae					
Plantago insularis				CM	
P. lanceolata				CM	
P. major				CM	
P. virginica				CM	
Order Rubiales					
Family Rubiaceae					
Galium stellatum var. *eremicum*				CM	
G. tinctorium				CM	
G. triflorum				CM	
Hedyotis pygmaea (= *Houstonia wrightii*)				CM	
Order Dipsacales					
Family Carpifoliaceae					
Sambucus glauca				CM	
S. microbotrys				CM	
Family Valerianaceae					
Valeriana capitata				CM	
V. edulis				CM	
V. occidentalis				CM	
Order Campanulales					
Family Campanulaceae					
Campanula parryi				CM	
C. rotundifolia				CM	
Lobelia anatina				CM	
L. cardinalis graminea				CM	
Order Asterales					
Family Compositae (= Asteraceae)					
Achillea millefolium var. *lanulosa* (= *A. lanulosa*)				CM	
Agoseris aurantiaca				CM	
A. glauca				CM	
Ambrosia psilostachya				CM	
Arnica chamissonis (= *A. foliosa*)				CM	
Artemisia biennis				CM	
A. tridentata				CM	
Aster adscendens				CM	
A. foliaceus				CM	
A. intricatus				CM	
A. spinosus				CM	
Baccharis emoryi				CM	
B. salicifolia (= *B. glutinosa*)				CM	
B. sarathroides				CM	
B. viminea				CM	
Bidens tenuisecta				CM	
Chrysothamnus nauseosus				CM	
Cichorium intybus				CM	
Cirsium nidulum				CM	
Cirsium sp.				CM	
Conyza canadensis (= *Erideron canadensis*)				CM	
Flaveria mcdougallii				CM	
Franseria confertifolia				CM	
Gnaphalium chilense				CM	
G. exilifolium (= *G. grayi*)				CM	
G. palustre				CM	

(continues)

TABLE 1.3.   *Continued*

Taxon[a]	Substrate[b] V	M,S	G,R	Source[c]	Feeding Habits[d]
Haplopappus acradenius				CM	
Helianthus ciliaris				CM	
Lactuca pulchella				CM	
L. serriola				CM	
Perityle emoryii				CM	
Solidago altissima				CM	
S. ana				CM	
S. decumbens				CM	
S. occidentalis				CM	
Sonchus asper				CM	
S. oleraceus				CM	
Stephanomeria pauciflora				CM	
Taraxcum officinale				CM	
Tessaria sericea (= *Plucea sericea*)				CM	
Verbesina enceloides				CM	
Xanthium strumarium (= *X. saccharatum*)				CM	
KINGDOM ANIMALIA					
Subkingdom Protozoa					
Class Mastigophora					
Order Cryptomonadida					
Family Cryptomonadidae					
Cryptomonas ovata, 1–27				CS	O
Order Chrysomonadida					
Family Ochromonadidae					
Dinobryon sertularia, 1–27		×	×	CS	O
Order Phytomonadida					
Family Carteridae					
Carteria klebsii, 1–27				CS	O
Family Chlamydomonidae					
Chlamydomonas sp., 1–27	×	×		CS	O
Family Volvacidae					
Pandorina morum, 1–27		×		CS	O
Order Euglenoidida					
Family Euglenidae					
Colacium sp., 1–27				CS	O
Trachelomas sp., 1–27				CS	O
Phylum Annelida					
Class Enchytraeidae, 1, 2, 8					
Class Hirudinoidae, 2, 12, 25					
Class Lumbricidae, 2, 12, 25					
Family Lumbriculidae					
Class Oligochaeta					
Order Haplotaxida					
Family Naididae					
Chaetogaster diaphanus, 1	×	×	×	BSS	C
Nais communis, 1	×	×	×	BSS	O
N. elinguis, 1, 2, 8; mile 51.6	×	×	×	BSS	O
N. pardalis, 1; mile 51.6	×	×	×	BSS	O
N. pseudobtusa, 2	×	×	×	BSS	O
N. variabilis, 1, 2, 8	×	×	×	BSS	O
Nais sp., 2; mile 51.6	×	×	×	BSS	O
Ophidonais serpentina, 1	×			BSS	O

TABLE 1.3. *Continued*

Taxon[a]	Substrate[b]			Source[c]	Feeding Habits[d]
	V	M,S	G,R		
Pristina sp., 1	×			BSS	O
Family Tubificidae					
Limnodrilus hoffmeisteri, 2, 12; mile 51.6		×		BSS	O
Tubifex tubifex, 1, 2; mile 51.6		×		BSS	O
Phylum Mollusca					
Order Basomatophora					
Class Gastropoda					
Family Cochlicopidae					
Cionella lubrica, 12, 18	×	×		SB	HD
Family Discidae					
Discus cronkhitei, 18	×	×		SB	HD
Family Helicarionidae					
Euconulus fulvus, 12	×	×		SB	HD
Family Helminthoglyptidae					
Sonorella coloradoensis, 12	×	×		SB	HD
Family Limacidae					
Deroceras laeve	×	×		SB	HD
Family Lymnaeidae					
Fossaria obtussa, 1, 2; 8, mile 51.6	×	×		SB	HD
F. parva	×	×		SB	HD
Family Oreohelicidae					
Oreohelix yavapai, 12	×	×		SB	HD
Family Planoboridae					
Gyraulus parvus		×		SB	HD
Family Physidae					
Physella humerosa, 12	×	×	×	SB	C
P. osculans, 12	×	×	×	SB	C
P. squalida, 12	×	×	×	SB	C
P. virgata virgata	×	×	×	SB	C
Physella sp., 1–25	×	×	×	SB	C
Family Pupillidae					
Gastrocopta ashmuni, 12	×	×		SB	HD
G. pelliculida	×	×		SB	HD
G. pilsbryana, 12	×	×		SB	HD
Pupilla blandi	×	×		SB	HD
P. hebes	×	×		SB	HD
P. syngenes	×	×		SB	HD
Pupilla sp.	×	×		SB	HD
Pupoides hordacea	×	×		SB	HD
P. nitidulus	×	×		SB	HD
Family Succineidae					
Catinella avara, 12, 13, 14, 18	×	×		SB	HD
Oxyloma haydeni kanabensis, 2	×	×		SB	HD
Succinea grosvenorii	×	×		SB	HD
Family Thysanophoridae					
Microphysula ingersolii	×	×		SB	HD
Thysanophora hornii	×	×		SB	HD
Family Valloniidae					
Vallonia cyclophorella, 12	×	×		SB	HD
V. perspectiva, 12	×	×		SB	HD
Family Vitrinidae					
Vitrina alaskana	×	×		SB	HD

(continues)

TABLE 1.3.   *Continued*

Taxon[a]	Substrate[b]			Source[c]	Feeding Habits[d]
	V	M,S	G,R		
Family Zonitidae					
Glyphyalinia indentata, 12	×	×		SB	HD
Hawaiia miniscula, 2	×	×		SB	HD
Zonitoides arborus	×	×		SB	HD
Order Veneroida					
Class Bivalvia					
Family Sphaeridae					
Pisidium variabile, 1		×		SB	HD
P. walkeri, 1					HD
Class Crustacea					
Subclass Cephalocardia					
Order Calanoida					
Family Diaptomidae					
Aglaodiaptomus clavipes		×		H	O
A. forbesi		×		H	O
Diaptomus sp.		×		H	O
Leptodiaptomus ashlandi		×		H	O
L. sicilis		×		H	O
Skistodiaptomus pallidus		×		H	O
S. reighardi		×		H	O
Order Cladocera					
Family Bosminidae					
Bosmina longirostris		×		H	OD
Family Chydoridae					
Alona affinis		×		H	OD
A. guttata		×		H	OD
Chydorus sphaericus		×		H	OD
Leydigia quadrangularis		×		H	OD
Pleuroxis aduncus		×		H	OD
P. denticulatus		×		H	OD
Family Daphnidae					
Daphnia galeata mendotae	×	×		H	OD
D. parvula	×	×		H	OD
D. pulex	×	×		H	OD
Daphnia sp.	×	×		H	OD
Diaphanosoma birgei	×	×		H	OD
Subclass Copepoda					
Order Cyclopodia					
Family Cyclopodae					
Acanthocyclops vernalis	×	×		H	OD
Diacyclops thomasi	×	×		H	OD
Eucyclops agilis	×	×		H	OD
E. speratus	×	×		H	OD
Mesocyclops edax	×	×		H	OD
Paracyclops fimbriatus poppei	×	×		H	OD
Tropocyclops prasinus mexicanus	×	×		H	OD
Order Harpacticoida					
Unident sp.		×		H	
Order Amphipoda					
Family Gammaridae					
Gammarus lacustris, 1–25	×	×		H	O
Subclass Ostracoda					
Order Podocopina					
Family Cypridae					

TABLE 1.3. *Continued*

Taxon[a]	Substrate[b] V	M,S	G,R	Source[c]	Feeding Habits[d]
Cypridopsis incongruens		×		H	O
C. pellucidus		×		H	O
C. salinus		×		H	O
C. vidua		×		H	O
Herpetocypris reptans	×	×	×	H	O
Ilyocypris bradyii				H	O
Paracandona euplectella				H	O
Potamocypris sp.				H	O
Class Insecta					
Order Diptera					
Family Chironomidae					
Apedilum subcinctum				BSS	
Cardiocladius platypus, 1, 8, 12, 21, 22; miles 51.6, 64.5, 183			×	BSS	C
Chironomus utahensis, 1	×	×		BSS	OH
Chironomus sp., 8, 12, 25; mile 51.6	×	×		BSS	OH
Cladotanytarsus sp., 8, 12, 25; 51.6 mile	×	×		BSS	O
Cricotopus annulator, 1–25	×	×	×	BSS	O
C. globistylus, 1, 12; 51.6 mile	×	×	×	BSS	O
C. infuscatus, miles 64.5, 193	×	×	×	BSS	O
C. trifascia, 8, 12, 20, 21, 22; miles 51.6, 64.5, 193	×	×	×	BSS	O
Cricotopus sp., 1, 2, 12, 20, 21, 22; mile 51.6	×	×	×	BSS	O
Cyphomella gibbera, 8, 12; mile 193				BSS	
Diamesa heteropus, 8			×	BSS	O
Eukiefferiella claripennis, 1–25	×	×	×	BSS	OC
E. coerulescens, 1–12	×	×	×	BSS	OC
E. devonica, 21, 22; mile 64.5	×	×	×	BSS	OC
E. ilkleyensis, 1–mile 193	×	×	×	BSS	OC
Eukiefferiella spp., 1	×	×	×	BSS	OC
Limnohyphes sp., 20	×	×	×	BSS	O
Metriocnemus sp., mile 51.6	×	×	×	BSS	OC
Micropsectra sp., 1	×	×		BSS	O
Orthocladius consobrinus, 20	×	×	×	BSS	HD
O. frigidus, 1	×	×	×	BSS	HD
O. luteipes, 2	×	×	×	BSS	HD
O. mallochi, mile 193	×	×	×	BSS	HD
O. rivicola, 1–25	×	×	×	BSS	HD
Parakiefferiella sp., 2, 12, 20	×	×	×	BSS	O
Paraphaenocladius exagitans, 3	×	×	×	BSS	O
Phaenopsectra profusa, 8, 21, 22, 25; mile 51.6	×	×		BSS	O
Polypedilum apicatum, mile 193	×			BSS	O
P. obelos, 8, 12, 21, 22; miles 51.6 193	×			BSS	O
Pseudosmitta sp., 2			×	BSS	O
Rheotanytarsus spp., 20			×	BSS	O
Tvetenia discoloripes				BSS	O
Family Simuliidae					
Simulium arcticum, 1–25			×	BSS	O
S. argus, 20, 21, 22; mile 193			×	BSS	O
S. griseum, 25			×	BSS	O
S. petersoni, 25			×	BSS	O

(continues)

TABLE 1.3.   *Continued*

Taxon[a]	V	M,S	G,R	Source[c]	Feeding Habits[d]
S. vittatum, 1			×	BSS	O
Family Tipulidae					
Order Trichoptera					
Order Hemiptera					
Family Corixidae					
Graptocorixa serrulata	×	×		PP	C
Family Gelastocoridae					
Gelastocoris oculatus	×	×		PP	C
Family Gerridae					
Gerris remigis		×	×	PP	C
Family Hebridae					
Hebrus hubbardi	×			PP	C
H. obscura	×			PP	C
Family Macroveliidae					
Macrovelia hornii		×	×	PP	C
Family Notonectidae					
Notonecta lobata	×	×		PP	C
Family Ochteriae					
Ochterus barberi	×		×	PP	C
O. rotundus	×		×	PP	C
Family Saldidae					
Saldula pallipes		×	×	PP	C
S. pexa		×	×	PP	C
Family Veliidae					
Microvelia beameri		×	×	PP	C
M. torquata		×	×	PP	C
Rhagovelia distincta			×	PP	C
Phylum Chordata					
Subphylum Vertebrata					
Class Osteichthyes					
Order Cypriniformes					
Family Catostomidae					
Catostomus latipinnis, widespread		×	×	CBT	O
Pantosteus (Catostomus) discobolus, widespread		×	×	CBT	O
Xyrauchen texanus, 1		×	×	CBT	O
Order Cypriniformes					
Family Cyprinidae					
Cyprinus carpio, widespread	×	×		CBT	O
Gila cypha, 8	×	×	×	CBT	O
G. elegans	×	×	×	CBT	O
G. robusta	×	×	×	CBT	O
Lepidomeda mollispinis				CBT	O
Notemigonus crysoleucus	×	×		CBT	C
Notropis lutrensis		×		CBT	O
Pimephales promelas, widespread		×		CBT	O
Plagopterus argentissimus				CBT	
Ptychocheilus lucius		×		CBT	C
Rhinichthys osculus, widespread		×		CBT	O
Order Salmoniformes					
Family Salmonidae					
Oncorhynchus kisutch			×	CBT	C
Salmo clarki			×	CBT	C

TABLE 1.3. *Continued*

Taxon[a]	Substrate[b]			Source[c]	Feeding Habits[d]
	V	M,S	G,R		
S. gairdneri, widespread			×	CBT	C
S. trutta, widespread			×	CBT	C
Salvelinus fontinalis, widespread			×	CBT	C
Order Siluriformes					
Family Ictaluridae					
Ictalurus melas		×		CBT	O
I. punctatus, 8, 20		×		CBT	O
Order Cyprinodontiformes					
Family Cyprinodontidae					
Fundulus zebrinus, 10, 20, Royal Arch		×		CBT	O
Order Perciformes					
Family Centrarchidae					
Micropterus salmoides	×	×		CBT	C
Chaenobryttus (= Lepomis) cyanellus	×	×		CBT	C
Lepomis macrochirus	×	×	×	CBT	O
Family Percichthidae					
Morone saxatilis			×	CBT	C
Family Percidae					
Stizostedion vitreum	×	×	×	CBT	C

[a]An asterisk indicates an introduced species. Numbers indicate the sites where the samples were collected.

[b]V, vegetation and debris; M, mud; S, sand; G, gravel; R, rubble.

[c]BSS, Blinn et al. (1992); CB, Czarnecki and Blinn (1978); CBT, Czarnecki et al. (1976); CM, Carothers and Minckley (1981b); CS, Crayton and Sommerfeld (1978); H, Haury (1986); PP, Polhemus and Polhemus (1976); SB, Spamer and Bogan (1993).

[d]Feeding habits: C = carnivore; HD = herbivore-detritivore; O = omnivore; OC = omnivore-carnivore; OD = omnivore-detritivore; OH = omnivore-herbivore.

boulders at Nankoweap (between river miles 51 and 52) because there was intense sandblasting of the boulders by sand carried in the strong current. The increase in *Cladophora glomerata* seemed to be correlated with channel depth at sites above Lees Ferry. Similarly, there was a decrease in biomass with depth at sites below Lees Ferry. The decrease in *Cladophora glomerata* at greater depths of the channel may have been due largely to the rapid attenuation of light due to periodically high sediment loads.

Large standing crops of *Cladophora glomerata* were found to be a habitat for the amphipod *Gammarus lacustris* and other small invertebrates. It also provided an enormous surface area for the attachment of epiphytic diatoms, which are an important food for aquatic invertebrates and in some cases for fish. Two hundred and thirty-five periphytic diatoms were reported by Czarnecki and Blinn (1978) in the seeps, in the mouths of tributaries, and in the Colorado River in the Grand Canyon reach. The greatest abundance of attached algae occurred in the summer months from June and July. Of the total number of taxa, 65% were diatoms, 24% cyanobacteria, and 10% chlorophytes. At the confluence of Diamond Creek a red algae, *Batrachospermum* sp., was reported. Another red alga that was present was *Audouienella* sp., often attached to the filaments of *Cladophora glomerata* and found in the deeper sections of the river channel.

Two hundred and thirty-five diatom taxa were also found in this reach of the river between the Marble and Grand Canyon systems. Here the dominant taxa were *Diatoma vulgare*, *Synedra ulna*, and *Cocconeis pediculus*. *Diatoma vulgare* and *Synedra ulna* were present in about the same amount before the impoundment. However, *Cocconeis pediculus* has increased in relative importance since impoundment. Their increase may indicate that *Cladophora glomerata* has also increased since the closure of Glen Canyon Dam. Usher et al. (1987) found that *Achnanthes affinis*, *Cocconeis pediculus*, *Diatoma vulgare*, and *Rhoicosphenia curvata* made up 80% of the communities upstream from Lees Ferry, whereas these four taxa were much less important at downstream sites, and *Gomphonema olivaceum*, *Cymbella affinis*, and *Nitzschia dissipata* became more important at the downstream sites. It may be that the change in species composition was due to their tolerance to suspended sediments (Bahls et al., 1984; Lowe, 1974).

Below Lees Ferry there was a fourfold decrease in epiphytic diatoms compared with sites above Lees Ferry, which corresponds to a decrease in cell density with increasing channel depth. The two patterns of diatom occurrence may both relate to the amount of suspended sediments, water depth, and light attenuation.

Invertebrates

It is difficult to compare the aquatic invertebrates before and after dam closure because of the introduction of many individuals of insects, snails, and leeches. Zooplankton and the algae seemed to be derived from lentic populations in Lake Powell. Haury (1986) proposed that occasional releases of surface water from spillways would enhance river populations, and nocturnal releases would have the greatest influence. Cladocera and copepods would rise temporarily due to the pattern of releases. There is evidence that the microcrustaceans below the dam increased to at least the mouth of Diamond Creek about 388 km below Glen Canyon Dam. However, Haury notes that the percentage of copepod planktors in poor condition increased downstream. Of the 34 invertebrates listed by Haury, only 16 were true planktors. The others were benthic. However, they sometimes break loose and drift downstream and are sampled in the plankton. These drifters contributed most of the biomass. Other planktors listed by Kubly (1976) were rotifers, a collembolan, and water mites. The last two are characterized best as epineustonic. The mites are probably nectonic. We can only speculate as to the origin of some of the crustacean planktors. *Aglaodiaptomus clavipes* and *Leptodiaptomus sicilis* were found in Lake Mead.

The backwaters that occurred along the borders of the river channel and at the mouths of tributaries are refugia for a rich fauna of invertebrates. There seemed to be a higher productivity of these invertebrates in the backwaters. The viability of these backwaters as habitats in the main stream of the Colorado River was also noted during a 1987–1989 survey (Kubly, 1976). Kubly noted a greater abundance in the quiet backwaters than in the main stream. The relative abundance of the Cladocera may be due to the preference of habitat but may also be due to the relative amount of fish predation.

Most sites for aquatic life are located near the mouths of tributaries and around debris caught in the river at the Grand Canyon National Park and vicinity. Just below the Glen Canyon Dam, the current is stochastic and the water is generally clear and deep. The rapid flow inhibits many types of organisms. Debris caught within the channel of the Grand Canyon and along the edges of the channel and at the mouths of tributaries furnish habitats for most of the aquatic life.

The vegetation ecosystem consists of algae and submerged leaves and stems of vascular plants—some mosses are also present. The Nostocaceae associated with this habitat were *Nostoc hatei, N. paludosum, N. punctiforme, N. verrucosum,* and unidentified species of *Nostoc.* Also present in this habitat was *Gloeotrichia intermedia,* which belongs to the family Oscillatoriaceae. Two algae that are members of the family Scytonemataceae, *Scytonema alatum* and *S. rivulare,* were also present; as was one member of the family Chlorococcaceae, species belonging to the genus *Chlorococcum*; and four taxons belonging to the family Hydrodictyaceae, all belonging to the genus *Pediastrum.* Other algae present were two taxons belonging to the genus *Tetraspora* of the order Tetrasporales. The green algae belonging to the order Ulotrichales were represented by one taxon belonging to the genus *Microspora* and seven taxons belonging to the genus *Ulothrix.* The family *Oedogoniaceae* was represented by unidentified species of the genus *Oedogonium.* Other species of green algae found associated with vegetation were two taxons belonging to the genus *Cladophora* and two belonging to the genus *Rhizoclonium* (Table 1.3).

The desmids were represented by three taxons belonging to the genera *Closterium* and *Cosmarium.* The family Zygnemataceae was represented by species belonging to the genus *Mougeotia* and one species belonging to the genus *Mougeotiopsis.* Diatoms were commonly found associated with the vegetation. They lived in and among the filaments of other algae or associated with the leaves of higher plants. They were three species belonging to the genus *Cyclotella* and two species belonging to the genus *Melosira.* Also present was one species, *Biddulphia laevis,* belonging to the family Biddulphiaceae. The family Fragilariaceae was represented by three taxons belonging to the genus *Fragilaria.* Also present were *Meridion circulare* and six taxons belonging to the genus *Synedra.* The genus *Achnanthes* was represented in this habitat by 12 taxons, the genus *Cocconeis* by four taxons, and the genus *Rhoicosphenia* by one taxon. The family Cymbellaceae was represented by seven taxons belonging to the genus *Amphora,* and the genus *Cymbella* was represented by 21 taxons. These two genera typically grow attached to substrates and therefore were probably attached to the stems or leaves of submerged plants or to the filaments of algae. The family Gomphonemaceae typically grow attached to algae or stems or leaves of plants or mosses. Eleven taxons were found in this habitat in this reach of the river (Table 1.3).

Associated with the algal filaments, or with plant debris, or in some cases with submerged aquatic plants were five taxons belonging to the genus *Diploneis.* The genus *Navicula,* which is typically found enmeshed but sometimes loosely attached to the vegetative substrates, was represented by 14 taxons. Associated with the vegetation were three taxons belonging to the genus *Stauroneis* and two taxons belonging to the genus *Denticula.* Also present within the vegetation were several members of the family Nitzschiaceae: *Bacillaria paradoxa, Hantzschia amphioxys,* and *H. amphioxys* f. *capitata,* and seven taxons belonging to the genus *Nitzschia* (Table 1.3).

A green alga, *Cladophora glomerata,* has become established on the riverbed as a result of increased light penetration and greater stability of the riverbed (Dolan et al., 1974; Blinn and Cole, 1991). Large seasonal extremes in discharged sediment load and temperature seem to have been eliminated since construction of the Grand Canyon Dam.

The discharge is now regulated by regional power demands. The maximum is about 566 cm s^{-1} and the daily minimum is 130 cm s^{-1} with extremes ranging from 28 to 764 cm s^{-1}. Water entering the river from near the base of the dam is clear and cold, the temperature ranging from 15 to 60°F (−9.5 to 15.56°C). The river is cooler in summer and warmer in winter than in predam days.

Invertebrate Fauna

The main-stem invertebrates of the Colorado River were generally low in productivity (biomass and density per square meter) except possibly in the section between Glen Canyon Dam and the confluence of the Little Colorado River, where the sediment input was minimal. Here the estimates of density were several thousand individuals per square meter. In contrast, grab samples taken below the confluence with the Little Colorado River usually yielded only 5 to 10 individuals m^{-2}. These were blackflies, midges, and aquatic earthworms, which were collected in exposed gravel bars in the stream channel and along the margins of the river (Table 1.3).

The density of invertebrates seemed to be lower at the confluence of the tributaries. Insects were fewer in spring and summer. In the Paria River, Elves Chasm, and Kanab Creek, there was little similarity between upstream and stations at the confluence with the Colorado River. This may be related to the difference in substrate type between the confluence and the upstream sampling areas. The tributaries seem to be most similar to each other in the spring. Only Hermit Creek deviated from this pattern. Compared with tributaries and other riverine systems, the reduced productivity in the main stream of the Colorado River seems to be due to the cooler annual temperature and the greater depth, the current velocity, and sediment input.

The colder annual temperatures reduce growth rates and hence the number of generations produced per year. Because of the hypolimnion water released from Glen Canyon Dam, the temperature was stressed in the river for many miles downstream and affected the invertebrate communities, resulting in less productivity and diversity. Some of the tributaries, such as the Little Colorado River, carried a great deal of sediment into the Colorado River proper, and this reduced light penetration and hence primary and secondary productivity. The reach between the Glen Canyon Dam and the Little Colorado River has the least input of sediments and hence has an increase in invertebrate productivity and diversity.

Common invertebrates in the main stream were freshwater amphipods and aquatic earthworms. These species were able to exist and reproduce in these habitats, whereas this was not possible with other invertebrates. In general the invertebrate biomass and density in the Colorado tributaries were substantially less than those in other streams in the Southwest (Oberlin et al., 1999). Several factors seemed to contribute to the low productivity of these tributaries. The high canyon walls to the north and south and the topographic variability combined to create a formidable dispersal barrier for many aquatic species. The canyon topography has effectively blocked the northward dispersal of some species that are present 128 km south of Oak Creek Canyon. As a result, the aquatic Hemiptera fauna, like other invertebrate orders in the canyon, is typical of the Southwest, although it is depaupered compared with collections made in other areas of Arizona that are similar to the canyon in habitat diversity and topography. Emergent forms of many aquatic insects are short-lived and wingless or poor flyers, such as the mayflies, stoneflies, dragonflies, damselflies, and water striders, which further reduces their dispersal ability.

Productivity and diversity were lowest in the spring and summer. It is probable that the spring runoff and summer flash floods from the tributaries disrupt the benthic invertebrate communities by washing out many species. This probably results in the similarity of the fauna in the tributaries. Flooding also reduces the algae populations on which many invertebrates depend. The small amount of flooding in fall and winter favored increased algae and invertebrate productivity in these seasons. Inactivity may also bring about a decrease in summer invertebrate populations in the tributaries. This is because the substrate is unstable, caused by streambed alterations, and thus would increase the drift of the insects. The water levels are high in the main stream of the Colorado River when the water at the confluence of each tributary is more similar to that of the main stream with regard to temperature, chemistry, suspended solids, sediments, and discharge, thus increasing the similarity between the fauna of the tributary and the main stream. The main stream of the Colorado River is typically cooler than that of the tributaries, and this may have a great influence on the invertebrate fauna. Furthermore, turbidity and suspended solids were higher in the main stream than in the tributaries: 986 mg L^{-1} and 170 JTU (Jackson turbidity units), respectively, in the main stream; 100 mg L^{-1} and 30 JTU in the tributaries (Cole and Kubly, 1976).

It is undoubtedly a combination of unstable physical and chemical parameters that reduces the fauna in the mouths of the tributaries and in the Colorado River just below the entrance of the tributaries. In general, the benthic invertebrate productivity in the Colorado River was lower than in the tributaries and in other rivers in the Southwest. It is believed to be due to the river's cooler annual temperature, greater depth, current velocity, and sediment input. In general, the areas that supported the richest invertebrate fauna were in rare shallow backwaters, eddies, along the margins of the river, and in the river section between Glen Canyon Dam and the Little Colorado River. Invertebrate groups found in the main stem were also present in various habitats in the tributaries. In general, the intratributary comparisons of the benthic invertebrate communities show that productivity and diversity were generally lower at the confluence. The physical and chemical effects produced by diurnal water fluctuations of the Colorado River on invertebrate fauna at the confluence appear to be substantial. Also, the instability of the substrate contributed to reducing the invertebrate fauna in diversity and productivity.

Fish Fauna

The fish fauna of the Colorado River has been studied mainly in the region known as the Grand Canyon and in Lake Powell and Lake Mead and areas just downstream from Lake Mead (Minckley, 1991b). The Colorado River Basin contains a mixture of native and introduced species. Minckley (1991b) states that although the families and species of fish in the Colorado River are relatively few, over 78% of the species of fish now known from the basin are particular to it. This is a larger percentage of species particular to a given river than is found in any other river in the United States. Because of long-term isolation and its unusual physical characteristics, the Colorado Basin supports one of the most distinctive ichthyofaunas in North America (Minckley, 1991b).

There are several distinctive features about the fish fauna of the Colorado River. The larger fish seem to live longer than in other rivers, and they are more streamlined and fusiform than most fish. Many have small depressed skulls, large predorsal humps or keels, or both, and elongated pencil-thin caudal peduncles may be present. Typically, they have small eyes, expansive and falcated (cycle-shaped) fins, and thick leathery skin. Scales are thin and deeply embedded in the skin and sometimes they almost seem to be absent. These special characteristics seem to be adapted to the severe habitats of the Colorado River. For example, hydrodynamic adaptation such as body and fin shape and the structure and size of the fish help it to maintain a position and maneuver itself in swift turbulent currents. To maintain themselves in the swift currents, large bodies and fins may be necessary. To minimize friction and counteract the abrasive sediments and provide a rigid external sheath to maximize muscle efficiencies, these fish often have small or reduced scales and leathery skin. Small eyes also help to reduce abrasive forces on eye structure. Long life seems to be desirable in this type of environment since the ability to reproduce may span decades rather than years in such an unpredictable habitat. Their large size does not seem adaptive to low discharge.

Most of the species of fish found in the Colorado River were described by 1900. Those collections made by Hubbs and Miller in the late 1930s validate earlier collections. It was quite evident that construction of the Hoover Dam changed the downstream fauna of fish. Some believed that the dams caused the reduction of native fish. Species preferring lentic habitats were introduced into the Colorado River after building of the dams. There was an introduction of bait fish after the dams were built and sport fishing increased (Miller, 1952). By 1960 the native fish population of the lower Colorado had been largely replaced by exotic species, due to the change of habitats and general characteristics of the water. In 1966, the Endangered Species Protection Act, which eventually turned into the Endangered Species Act of 1973, reduced the poisoning and the introduction of undesirable fish.

The National Environmental Protection Act of 1969 recognized that habitats were being lost at an unacceptable rate and mandated assessments and disclosures of impacts of federal projects. Minckley (1991b) lists eight native species of fish in the Grand Canyon National Park, six of which are endemic. The speckled dace and the roundtailed chub are known from adjacent rivers. The humpback chub, bonytail, Colorado spikeminnow, and razorback suckers are listed or proposed as endangered by the Department of Interior (U.S. Fish and Wildlife Service, 1988). Of these species, only the humpback chub has reproductive populations. Other fish are extremely depleted or extirpated. Speckled dace, flannelmouth suckers, and blue suckers remain relatively common (Minckley, 1985).

From Glen and Marble Canyons, 17 species were collected at 27 sites. Most of these were from tributaries. The roundtailed chub, the Colorado spikeminnow; the speckled dace; and flannelmouth, bluehead, and razorback suckers were the native fish that were collected. The nonnative species were the flathead minnow, carp, channel catfish, and green sunfish. The humpback chub was described from three specimens, two from an unknown creek and one from Bright Angel Creek.

Near Lees Ferry the following fish were found: speckled dace, spikeminnow, flannelmouth, bluehead suckers, and nonnative channel catfish in 1934. In Nankoweap Creek dace were caught (Lowe, 1974). A humpback chub was caught in Spencer Creek by Wallace in 1955. After closure of the Glen Canyon Dam, the only large species were the humpback and roundtailed chub, bonytail, and some hybrids (Holden, 1968; Stone and Rathbon, 1967, 1969). Introduced carp, channel catfish, flannelmouth suckers, and bluehead were also there.

The fish fauna is best known between Lees Ferry and Separation Rapids. The species in this reach are the bonytail chub (*Gila elegans* rare), humpback chub (*G. cypha*), the Colorado River chub (*G. robusta*), *Ptychocheilus lucius*, speckled dace (*Rhinichthys osculus*), razorback sucker (*Xyrauchen texanus*), the flannelmouth sucker (*Catostomus latipinnis*), and the bluehead sucker (*Pantosteus* (*Catostomus*) *discobulus*). Introduced species are Coho salmon (*Oncorhynchus kisutch*), rainbow trout (*Salmo gairdneri*), cutthroat trout (*S. clarki*), brown trout (*S. trutta*), brook trout (*Salvelinus fontinalis*), carp (*Cyprinus carpio*), golden shiner (*Notemigonus crysoleucus*), Virgin River spine dace (*Lepidomeda mollispinis*), woodfin (*Plagopterus argentissimus*), red shiner (*Notropis lutrensis*), fathead minnow (*Pimephales promelas*), channel catfish (*Ictalurus punctatus*), black bullhead (*I. melas*), Rio Grande killifish (*Fundulus zebrinus*), striped bass (*Morone saxatilis*), largemouth bass (*Micropterus salmoides*), green sunfish (*Chaenobryttus cyanellus*), bluegill sunfish (*Lepomis macrochirus*), and the walleye (*Stizostedion vitreum*) (Table 1.3).

Twenty-seven fish species are known from the Colorado River in the study area, of which 70% are exotics. In this reach of the Colorado River the fish fauna is composed of species that have been introduced and also a few native species. Several of the native species have become extinct.

Originally in the Colorado River there were eight native species of fish, including the bonytail chub, Colorado (roundtail chub), and Colorado spikeminnow, which are apparently extinct in the Grand Canyon. This disappearance is not restricted to the Grand Canyon area alone since all three have been designated endangered species. Another endangered species native to the Colorado River system is the humpback chub. Populations have been markedly reduced since major habitat changes, although they are still extant in the Grand Canyon area. Recently, it reproduces only near the confluence of the Little Colorado and Colorado Rivers in the canyon. The razorback sucker is thought to be going extinct throughout its former range within the Colorado River basin. It was considered extinct in the Grand

Canyon until three adults were captured near the confluence of the Paria and Colorado Rivers. In the late 1970s and early 1980s, the three native species—the bluehead and flannelmouth suckers and the speckled dace—were represented by what appeared to be healthy reproducing populations through most of their former range. Juvenile suckers remained in the perennial tributaries for two to three years after hatching before moving into the main stream. The speckled dace are found regularly in the tributaries, although their occurrence in the main stream is highly variable.

Nineteen exotic species have been introduced into the Colorado River system. Ten of the 19 exotic species have been observed so infrequently that they should be considered an insignificant component of the Grand Canyon ichthyofauna. Exotic species or introduced species that have been recorded from this area are as follows: golden shiner, green sunfish, bluegill sunfish, Coho salmon, black bullheads, largemouth bass, and red shiner. Largemouth bass and bluegills have occasionally been recorded from the Grand Canyon area. The bass seem to be very localized in occurrence.

The bass was suprisingly the most common species encountered from 237 Mile Rapids to Separation Canyon. However, there are undocumented accounts of striped bass occurring sporadically upstream at Havasu river mile 156.9. Two exotic minnow species that seem to be well established in the Grand Canyon are the fathead minnow and the Rio Grande killifish. They seem to have reached the highest density in the Little Colorado River and Unkar and Kanab Creeks. The four remaining exotic species are trout: rainbow, brown, brook, and cutthroat. All of these have been introduced in sports fishery stocking programs. Rainbow trout is most widely distributed in the area. The brown and brook trout are far less numerous. Cutthroat trout were introduced at Lees Ferry. Capture was in November 1979. It was the most common fish taken by anglers at Badger Rapids.

Although channel catfish are hard to catch, they are probably ubiquitous in the study area. Their density is low above Lava Falls and considerably higher from Lava Falls (between river miles 179 and 180 downstream) with a definite higher concentration occurring between Separation Rapids and the rapids 5 km upstream (between river miles 239 and 240). These fish are fairly common in this reach but are most often found near the confluence of the Little Colorado River. Carp was the most common exotic species found in the Grand Canyon in the collections of Carothers and Minckley.

Twenty-seven species of fish are known to be present or have occurred in the Colorado River and its tributaries in the Grand Canyon reach. The populations of these species are very variable. They rank in abundance in this area as follows: carp, speckled dace, flannelmouth sucker, rainbow trout, bluehead sucker, and humpback chub. Brown and brook trout, fathead minnows, channel catfish, and the Rio Grande killifish are the only other species that appear to maintain stable but low-density populations in the Grand Canyon area.

The juvenile humpback chub forages on or near the substrate, selecting benthic insect larvae and organic detritus as its primary food items. For example, food items identified in the stomach of a single winter mortality consisted of several midge larvae (Chironomidae) and a biting midge larve (Ceratopogonids). Specimens found below Glen Canyon Dam were feeding on planktonic crustacea, which apparently came from Lake Powell. Roundtail chub (*Gila robusta*) is omnivorous, feeding on aquatic and terrestrial insects as well as filamentous algae. *Ptychocheilus lucius*, which is certainly endangered if not extinct in the Grand Canyon, eats primarily larvae, pupae, and nymphs of aquatic insects as well as cladocerans and copepods.

Speckled dace (*Rhinichthys osculus*) feeds on aquatic invertebrates, especially Ephemeroptera, Diptera, and Trichoptera. The blackfly and midge larvae are the most common. Other major food items included mayfly nymphs in winter, seed shrimp (Cypridae) in summer, and net-spinning caddisfly larvae in summer and fall. Larvae of Dobson flies and soldier flies were secondary in importance during the spring. Terrestrial insects were a minor component of the seasonal diets.

In the northern tributaries the speckled dace fed primarily on benthic insects and amorphous organic debris. Adult midges and ants were utilized in small to moderate quantities throughout the year. Other terrestrial insects eaten by the speckled dace included leafhoppers during the winter, a plant hopper, an ichneumonid wasp in the spring, and a ctenuchid moth in the fall. In the southern tributaries the food was similar to that in northern tributaries. Benthic invertebrates predominated in the diet, with blackfly and midge larvae being the most heavily exploited taxa. Other major food items included mayfly nymphs in winter and seed shrimp in summer and net-spinning caddisfly larvae in summer and fall. As in the Grand Canyon area larvae of Dobson flies and soldier flies were of secondary importance in the spring. Terrestrial insects were a minor component of the seasonal diet, with pyralid and ctenuchid moths, adult midges, and ants being the only taxa comprising greater than 0.5% of any seasonal diet. Nonanimal diets included the algae *Nostoc* and detritus. *Nostoc* was eaten primarily through the winter and was limited primarily to speckled dace collected from Hermit Creek.

The razorback sucker is an omnivorous species. The food includes algae and dipteran larvae and may include crustaceans, particularly in the month of May. The flannelmouth sucker is omnivorous and feeds heavily on midges, blackflies, scuds, organic debris, and *Cladophora*. The bluehead sucker (*Catostomus discobolus*) feeds extensively on invertebrates in the study area, their diets consisting primarily of dipterans and scuds. In the main stem of the Colorado River, they fed mostly on blackfly larvae and midge larvae. Other aquatic invertebrates consumed included cranefly larvae in winter and net-winged midge larvae in summer. Adult blackflies and midges were present in small quantities during the summer. Invertebrate gut content included organic debris, which was the principal dietary component during most of the year, and *Cladophora*. Digestive tracts of the bluehead sucker consisted of abundant diatoms of all seasons. Winter samples were dominated by *Diatoma*, *Cymbella*, and *Gomphonema*, whereas *Diatoma*, *Navicula*, *Rhoicosphenia*, and *Cocconeis* predominated during the summer and fall, and in the fall, mainly *Diatoma* and *Cocconeis*.

The main food of rainbow trout (*Salmo gairdneri*) in the main stem of the Colorado were aquatic invertebrates, primarily scuds and blackfly larvae during winter and fall and terrestrial insects in spring and summer. *Cladophora* was an important dietary item throughout the year. Rainbow trout in tributaries utilized benthic invertebrates, *Cladophora*, and to a lesser extent than in the main stem, terrestrial insects. In the main stem the rainbow trout consumed 65 different food items throughout the year. The riverine diet of this species reflects its visual orientation, large body size, and ability to actively exploit food resources throughout its habitat. *Cladophora* constituted from one-fourth to one-half of the dietary intake throughout the year. Based on invertebrate taxa identified from algal masses in the stomach, *Cladophora* was taken from both the substrate and drift. Invertebrates constituted an important element in the annual diet of the main-stem rainbow trout. Aquatic insects dominated the diet in the fall and winter, while terrestrial insects were more heavily utilized in the spring and summer. Scuds comprised the major invertebrate prey of the rainbow trout. Small amounts of midge larvae and pupae were consumed throughout the year. The most heavily exploited terrestrial insects were stinkbugs, grasshoppers, ants, and scarab beetles, which collectively composed a considerable amount of the spring and summer diet.

In the northern tributaries, 61 types of food were ingested by the rainbow trout and the diet was similar to that of the main-stream individuals. *Cladophora* constituted the major portion of the diet in all seasons except summer, when it represented only 8.32% of the stomach contents. Dobson fly larvae appeared to be utilized only in the winter and summer. Individual families of caddisflies were ingested in large quantities whenever present. Net-spinning caddisflies were eaten during the winter and summer, small-case caddisflies (Heliopsychidae) and northern caddisflies (Limnephilidae) in the spring, and fingernet (Philopotamidae) and microcaddisflies in the fall. Collectively, these three families formed from 0.31 to 7.02% of each season's diet. Dipteran pupae were utilized only slightly throughout the year (0.02 to 0.89%), while the larvae, especially of blackflies and midges, were utilized extensively in the summer (25.79%). Scuds were identified in the stomach contents from three of the four sampling periods. Feeding on terrestrial insects was of minor importance in northern tributaries, as terrestrial insects made up only 1.28 to 5.09% of the spring and summer diets, respectively.

Forty-six types of food were identified in the stomachs of rainbow trout taken from southern tributaries. The food was strikingly different in these tributaries from that in the main stream and northern tributaries. *Cladophora* was present in small quantities during fall and winter and absent from the diet in spring and summer. Mayfly nymphs were the major prey item throughout much of the year, ranging from 1.9 to 24.63% of the total diet. Pouch snails (Physidae) were the most common aquatic invertebrate present in the stomach contents during the summer (18.18%) and fall (36.52%). Common Diptera in the diet included blackfly, soldier fly, midge, horsefly, and moth fly larvae, as well as midge and blackfly pupae. Together, these taxa made up 3.48 to 28.29% of the seasonal diet. Dobson fly larvae were heavily preyed upon only in the fall (15.89%). Other aquatic invertebrates consumed in more than trace amounts were predacious diving beetles (Dytiscidae), seed shrimp, net-spinning caddisflies, and scuds. Terrestrial insects, especially grasshoppers, blackflies, and bees, were common prey items in the summer (10.99%) and fall (12.51%).

For brown trout (*Salmo trutta*), scuds and immature dipterans were the major food items throughout the year. However, no specimens were collected in the winter. Several groups, such as pentatomids, ctenuchids, and *Cladophora,* were important seasonally. Aquatic organisms in the stomachs of main-stem river brown trout consisted primarily of scuds, blackfly larvae, and pupae and *Cladophora*. Dominant terrestrial insects in order of importance were stinkbugs, ctenuchid moths, ants, adult blackflies, and grasshoppers. Seasonal variation in brown trout diet were similar to those of rainbow trout. Specimens taken in the spring contain large numbers of *Cladophora*, terrestrial insects, and scuds, which collectively formed 82.3% of the diet. Blackfly larvae and pupae were the primary summer food items and accounted for 64.15% of the total diet. The fall diet consisted primarily of scuds (52.12%), immature blackflies (16.68%), and terrestrial insects (22.24%).

Brook trout (*Salvelinus fontinalis*) feed primarily on larger aquatic invertebrates: scuds, earthworms, and snails. Also present were *Cladophora*, midge pupae, blackfly larvae and pupae, midge larvae, and ants. Adult midges, grass seeds, aphids, and leafhoppers were present in small amounts.

Carp (*Cyprinus carpio*) feeds primarily on *Cladophora* during all seasons. Scuds and midge larvae are the most commonly ingested aquatic invertebrates. They comprise 0.09 to 11.12% of the seasonal diet, the lesser amount in winter and the larger amount in spring. Additional invertebrate prey included aquatic earthworms, blackfly larvae, and midge and blackfly pupae. Few terrestrial insects were ingested. Speckled dace was the only vertebrate found in the carp's stomach. *Cladophora*, which consistently made up the greatest portion of the seasonal diet (55.1 to 86.6%), was most heavily utilized in the summer and fall. Organic detritus and plant seeds were also major dietary items.

The golden shiner (*Notemigonus crysoleucus*) has as its main food blackfly larvae. The Virgin River spine dace (*Lepidomeda mollispinis*) fed primarily on aquatic insect larvae, although algae were also taken when aquatic invertebrates were not present. Woundfin (*Plagopterus argentissimus*), an endangered species, fed primarily on animal matter, although debris, plant material, and algae are also ingested. The red shiner (*Notropis lutrensis*) feeds on plankton, benthic invertebrates, juvenile fish, terrestrial insects, and fish eggs. Fathead minnow (*Pimephales promelus*) fed on dipterans, organic detritus, and small amounts of filamentous algae. Channel catfish (*Ictalurus punctatus*) fed on scuds, immature dipterans, and *Cladophora*. It is truly an omnivore and feeds more on plant material in some of the tributaries. Black bullhead (*I. melas*) is an omnivore, although it becomes carnivorous when animal prey items are abundant. Midge larvae, mayflies, leafhoppers, and organic detritus were the major food items consumed by the Rio Grande killifish (*Fundulus zebrinus*). Flannelmouth bass (*Micropterus salmoides*) feeds largely on small invertebrates, especially zooplankton and aquatic insects. The stomach content of green sunfish (*Lepomis cyanellus*) taken from the main stream during the fall contained primarily blackflies and midges. Unidentified invertebrate remains were a considerable portion of the diet. The food of the young walleye pike (*Stizostedion vitreum*) consists mainly of small crustacea, insects, and fish, while the adults feed mainly on other fish.

STRUCTURE OF AQUATIC COMMUNITIES—
EXAMPLE: VASEY'S PARADISE ECOSYSTEM

Vegetative Communities

Vasey's Paradise enters the Colorado River in a reach where riverine conditions exist and the effects of flooding are evident (Figure 1.2). Associated vegetation were the green alga *Pediastrum duplex* var. *clathratum*. In and among the plants were filaments of *Ulothrix* sp. and *Mougeotiopsis* sp.; and the diatom *Cyclotella meneghiniana*, which was probably on the sediment or on the filaments of green algae. Also present was *Melosira varians*, which forms filaments and is probably attached to various plants in the area. Other diatoms present were *Fragilaria vaucheriae*, *Synedra tenera* var. *genuina*, *S. ulna*, and *S. ulna* var. *constricta*. Attached to the algae or other plant material in the area were *Achnanthes affinis*. Widespread were *Achnanthes lanceolata* and *A. lanceolata* var. *dubia*. Also widespread were *A. linearis* forma. *curvata* and *A. linearis* var. *pusilla*. *Achnanthes minutissima* was also very common, as was *Cocconeis pediculus*, which was probably attached to the filamentous algae or other plant stems or leaves that might be present. This was also true for *C. placentula*

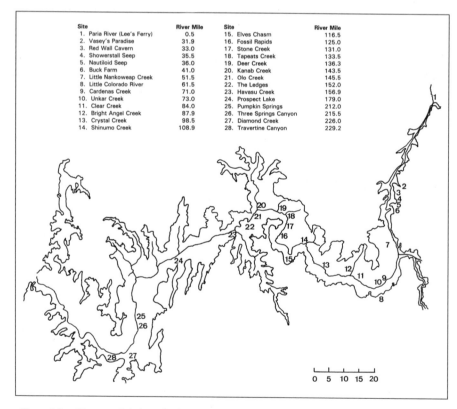

Site	River Mile	Site	River Mile
1. Paria River (Lee's Ferry)	0.5	15. Elves Chasm	116.5
2. Vasey's Paradise	31.9	16. Fossil Rapids	125.0
3. Red Wall Cavern	33.0	17. Stone Creek	131.0
4. Showerstall Seep	35.5	18. Tapeats Creek	133.5
5. Nautiloid Seep	36.0	19. Deer Creek	136.3
6. Buck Farm	41.0	20. Kanab Creek	143.5
7. Little Nankoweap Creek	51.5	21. Olo Creek	145.5
8. Little Colorado River	61.5	22. The Ledges	152.0
9. Cardenas Creek	71.0	23. Havasu Creek	156.9
10. Unkar Creek	73.0	24. Prospect Lake	179.0
11. Clear Creek	84.0	25. Pumpkin Springs	212.0
12. Bright Angel Creek	87.9	26. Three Springs Canyon	215.5
13. Crystal Creek	98.5	27. Diamond Creek	226.0
14. Shinumo Creek	108.9	28. Travertine Canyon	229.2

Figure 1.2.   Sites sampled along the Colorado River in the Grand Canyon National Park and vicinity.
(From Crayton and Sommerfeld, 1978; Stevens, 1983.)

var. *euglypta* and var. *lineata*. The family Cymbellaceae was very common associated with plant material. *Amphora ovalis* var. *pediculus* was widespread, as were *Cymbella affinis* and *C. amphicephala*. Other *Cymbella* species associated with plant debris and with the stems and leaves of plants or the filaments of plants were *Cymbella aspera*, *C. caespitosa* var. *ovata*, *C. hungarica*, and *C. incerta*. Widespread were *Cymbella laevis*, *C. minuta*, and *C. prostrata*. Other common diatoms associated with the vegetation were *Gomphoneis herculeana*, *Gomphonema parvulum*, *G. subclavatum*, and *G. truncatum*. Four taxons belonging to the genus *Diploneis* were also found (Table 1.4) as were three taxons of *Navicula cryptocephala*. These were very common in Vasey's springs. Other *Navicula* species present were *Navicula mutica*, which was widespread; *Navicula mutica* var. *cohnii*; and *N. radiosa*, which was widespread, as was its variety *tenella*. Other widespread *Navicula* species were *N. tripunctata*, *Denticula elegans*, and *D. rainerensis*. Also widespread within the vegetation were *Hantzschia amphioxys* and *Nitzschia apiculata*. Other widespread *Nitzschia* species were *N. dissipata* and *N. frustulum*. Thus the vegetation of both algae and other groups was commonly associated with diatoms.

Several protozoans were found in the habitats associated with the algae. They were an omnivore, *Dinobryon sertularia*, and a protozoan, *Chlamydomonas* sp., which is an omnivore. Other omnivores were the worms *Nais pseudobtusa* and *N. variabilis* and an unidentified species of *Nais*. Gastropods were found associated with plant material in the springs. They were an herbivore–detritivore, *Fossaria obtrussa*; a carnivore, *Physella* sp.; a herbivore–detritivore, *Oxyloma haydeni kanabensis*; and a herbivore–detritivore, *Hawaiia miniscula*. Within the vegetation was found an omnivorous arthropod, *Gammarus lacustris*. Certain chironomids were associated with the debris in Vasey's springs: the omnivorous *Cricotopus annulator* and an unidentified species of *Cricotopus*, which is probably an omnivore. An omnivore–carnivore chironomid, *Eukiefferiella claripennis*, was present, probably feeding on the protozoans and algae and other organisms. Also present was an omnivore–carnivore, *E. coerulescens*. Herbivore–detritivores were also present, as indicated by the occurrence of *Orthocladius luteipes* and *O. rivicola*. Only one fish was reported associated with the vegetation: *Cyprinus carpio*, an omnivore. It is probably an introduced species.

Lentic Communities: Mud and Sand

In muddy sand habitats formed on the bed of the springs were several algae: three taxons belonging to the blue-green algal genus *Oscillatoria*. Also present were three taxons belonging to the genus *Spirulina*. They were found lying on the mud and sand or sandy mud in this habitat. Another species found in this habitat was *Pediastrum duplex* var. *clathratum*. Also present were three taxons belonging to the genus *Scenedesmus*. Another green alga found was a desmid, *Staurastrum* sp.

Several diatoms seemed to prefer this habitat, that is, the mud or sand or muddy sand in this habitat: *Cyclotella meneghiniana* and *Melosira varians*, which were probably attached to or associated with debris in the sandy mud. Also associated with the sandy mud were *Fragilaria construens*; *F. vaucheriae*, which was common; *F. virescens*; and *Synedra acus*, which was probably attached to debris within the sandy mud. Other diatoms present were *Synedra socia*; *S. ulna*, which was widespread; and

(text continues on page 55)

TABLE 1.4.   *Species List: Vasey's Paradise in Grand Canyon National Park*

Taxon[a]	Substrate[b] V	M,S	G,R	Source[c]	Feeding Habits[d]
SUPERKINGDOM PROKARYOTAE					
KINGDOM MONERA					
Division Cyanophycota					
Class Cyanophyceae					
Order Chamaesiphonales					
Family Chamaesiphonaceae					
Chamaesiphon incrustans, 1–27				CS	
Order Chroococales					
Family Chroococaceae					
Chroococcus sp., 1–27				CS	
Order Nostocales					
Family Nostocaceae					
Nodularia spumigena, 1–27				CS	
Family Oscillatoriaceae					
Oscillatoria articulata, 1–27		×	×	CS	
O. limosa, 1–27		×	×	CS	
O. nigra, 1–27		×	×	CS	
Spirulina major, 1–27		×	×	CS	
S. subsalsa, 1–27		×	×	CS	
Spirulina sp., 1–27		×	×	CS	
SUPERKINGDOM EUKARYOTAE					
KINGDOM PLANTAE					
Subkingdom Thallobionta					
Division Chlorophycota					
Order Chaetophorales					
Family Chaetophoraceae					
Cloniophora sp., 1–27			×	CS	
Stigeoclonium flagelliferium, 1–27			×	CS	
Stigeoclonium sp., 1–27			×	CS	
Tetraedron sp., 1–27			×	CS	
Family Hydrodictyaceae					
Pediastrum duplex var. *clathratum*, 1–27	×	×		CS	
Family Oocystaceae					
Cerasterias staurastroides, 1–27				CS	
Lagerheimie subsalsa, 1–27				CS	
Family Scenedesmaceae					
Scenedesmus bijuga, 1–27		×		CS	
S. opoliensis, 1–27		×		CS	
S. quadricauda var. *maximus*, 1–27		×		CS	
Order Trentepohliales					
Family Trentepohliaceae					
Leptosira sp., 1–27					
Order Ulotrichales					
Family Microsporaceae					
Microspora floccosa, 1–27			×	CS	
Ulothrix sp., 1–27	×			CS	
Family Monostromataceae					
Gomontia sp., 1–27				CS	
Order Zygnematales					
Family Desmidiaceae					
Staurastrum sp., 1–27		×		CS	

TABLE 1.4. *Continued*

Taxon[a]	Substrate[b]			Source[c]	Feeding Habits[d]
	V	M,S	G,R		
Family Zygnemataceae					
Mougeotiopsis sp., 1–27	×		×	CS	
Class Bacillariophyceae					
Order Eupodiscales					
Family Coscinodiscaceae					
Cyclotella meneghiniana, throughout	×	×	×	CBT	
Colorado River					
Melosira varians, widespread	×	×	×	CBT	
Order Fragilariales					
Family Fragilariaceae					
Asterionella formosa, upper Colorado River			×	CBT	
Diatoma anceps, 1–27			×	CS	
D. elongatum, 1–27			×	CS	
D. hiemale, 1–27			×	CS	
D. hiemale var. *mesodon*, throughout			×	CB	
Glen Canyon, especially 18					
D. vulgare, widespread			×	CBT	
Fragilaria aequalis, 1–27				CS	
F. brevistriata, 1–27				CS	
F. construens, 1–27		×	×	CS	
F. intermedia, 1–27			×	CS	
F. vaucheriae, widespread	×	×	×	CBT	
F. virescens, 1–27		×		CS	
Synedra acus, common throughout		×	×	CBT	
Glen Canyon, abundant 13					
S. nana, 1–27		×		CS	
S. socia, widespread, high densities above		×	×	CBT	
Little Colorado River					
S. tenera var. *genuina*, 1–27	×		×	CS	
S. ulna, widespread	×	×	×	CBT	
S. ulna var. *constricta*, 2	×	×	×	CBT	
Order Achnanthales					
Family Achnanthaceae					
Achnanthes affinis, widespread	×	×	×	CBT	
A. exigua var. *heterovalva*, widespread,		×	×	CBT	
especially 2					
A. lanceolata, widespread in Glen Canyon	×	×	×		
A. lanceolata var. *dubia*, widespread in	×	×	×	CBT	
Glen Canyon					
A. linearis f. *curvata*, widespread in	×	×	×	CBT	
Glen Canyon					
A. linearis var. *pusilla*, widespread in	×	×	×	CBT	
Glen Canyon					
A. microcephala, widespread in Glen Canyon		×	×	CBT	
A. minutissima, widespread	×	×	×	CBT	
Achnanthidium sp., 1–27				CS	
Cocconeis diminuta, 2, 12, 15, 19			×	CBT	
C. pediculus, widespread	×		×	CBT	
C. placentula var. *euglypta*, widespread	×		×	CBT	
C. placentula var. *lineata*, widespread	×		×	CBT	
Rhoicosphenia curvata, widespread	×		×	CBT	
Order Naviculales					
Family Cymbellaceae					
Amphora ovalis var. *pediculus*, widespread	×		×	CBT	

(continues)

TABLE 1.4.   *Continued*

Taxon[a]	Substrate[b]			Source[c]	Feeding Habits[d]
	V	M,S	G,R		
Cymbella affinis, widespread	×		×	CBT	
C. amphicephala, widespread	×		×	CBT	
C. aspera, 1–27	×		×	CS	
C. caespitosa var. *ovata*, 1–27	×		×	CS	
C. hungarica, 1–27	×		×	CS	
C. incerta, 1–27	×		×	CS	
C. laevis, widespread in small numbers	×		×	CBT	
C. minuta, widespread	×		×	CBT	
C. prostrata, widespread	×	×		CBT	
C. tumida, widespread		×	×	CBT	
C. turgida, 1–27			×	CS	
C. ventricosa, 1–27		×	×	CS	
C. ventricosa var. *semicircularis*, 1–27		×	×	CS	
Family Gomphonemaceae					
Gomphoneis herculeana, widespread	×			CBT	
G. parvulum, widespread	×			CBT	
G. subclavatum, widespread	×			CBT	
G. truncatum, widespread	×			CBT	
Family Naviculaceae					
Amphipleura pellucida, widespread		×		CBT	
Anomoeoneis exilis, 1–27		×		CS	
A. serians var. *brachysira*, 1–27		×		CS	
A. vitrea, 11, 27		×		CBT	
Caloneis bacillum, widespread		×		CBT	
Diploneis elliptica, 2	×			CBT	
D. oblongella, 2	×			CBT	
D. oculata, 2, 8	×			CBT	
D. puella, widespread	×			CBT	
Frustulia vulgaris, widespread		×	×	CB	
Mastogloia smithii, widespread				CBT	
M. smithii var. *lacustris*, widespread				CBT	
Navicula angelica, 1–27		×		CS	
N. angelica var. *subsalsa*, widespread		×		CBT	
N. arvenensis, widespread		×		CBT	
N. cryptocephala, widespread	×	×	×	CBT	
N. cryptocephala f. *minuta*, widespread	×	×	×	CBT	
N. cryptocephala var. *veneta*, widespread	×	×	×	CBT	
N. decussis, widespread		×	×	CBT	
N. graciliodes, widespread		×		CB	
N. minima, widespread			×	CBT	
N. mutica, widespread	×	×	×	CBT	
N. mutica var. *cohnii*, 2, 15, 18, 27	×	×	×	CBT	
N. pupula var. *rectangularis*, widespread		×		CBT	
N. radiosa, widespread	×	×		CBT	
N. radiosa var. *tenella*, widespread	×	×		CBT	
N. subtilissima, 2, 6		×		CBT	
N. tripunctata, widespread	×		×	CBT	
N. zanoni, widespread, most common in 2, 14, 18				CBT	
Neidium dubium, 1–27		×	×	CS	
Pinnularia sp., 1–27				CS	
Rhopalodia gibba, widespread				CBT	

TABLE 1.4. *Continued*

Taxon[a]	Substrate[b]			Source[c]	Feeding Habits[d]
	V	M,S	G,R		
R. gibberula var. *vanheurckii*, widespread				CBT	
Order Epithemiales					
Family Epithemiaceae					
Denticula elegans, widespread	×		×	CBT	
D. rainerensis, 2, 4, others	×		×	CBT	
Epithemia adnata, 2, 6				CBT	
E. argus var. *alpestris*, 2, 6				CBT	
E. argus var. *longicornis*, widespread				CBT	
E. sorex, widespread				CBT	
E. turgida, upper reaches of Colorado River, below Glen Canyon, but not in Lake Powell area				CBT	
Order Bacillariales					
Family Nitzsciaceae					
Hantzschia amphioxys, widespread	×			CBT	
Nitzschia acicularis, widespread		×		CBT	
N. amphibia, widespread			×	CBT	
N. apiculata, widespread	×			CBT	
N. communis, widespread				CBT	
N. denticula, 2				CBT	
N. dissipata, widespread	×	×	×	CBT	
N. frustulum, widespread	×	×	×	CBT	
N. kutzingiana, widespread			×	CBT	
N. linearis, widespread				CBT	
N. romano, 2, 12, 15		×		CM	
Order Surirellales					
Family Surirellalaceae					
Campylodiscus noricus var. *hibernica*, 2				CBT	
Surirella angustata, widespread		×		CBT	
S. brightwellei, widespread		×		CBT	
Division Chromophycota					
Class Xanthophyceae					
Tetragoniella sp., 1–27				CS	
Order Vaucheriales					
Family Tribonemataceae					
Tribonema utriculosum, 1–27				CS	
Class Dinophyceae					
Order Peridinales					
Family Ceratiaceae					
Ceratium carolinianum, 1–27			×	CS	
C. hirudinella, 1–27			×	CS	
KINGDOM ANIMALIA					
Subkingdom Protozoa					
Class Mastigophora					
Order Cryptomonadida					
Family Cryptomonadidae					
Cryptomonas ovata, 1–27		×		CS	O
Order Chrysomonadida					
Family Ochromonadidae					
Dinobryon sertularia, 1–27	×	×		CS	O
Order Phytomonadida					
Family Carteridae					

(continues)

TABLE 1.4.   *Continued*

Taxon[a]	Substrate[b]			Source[c]	Feeding Habits[d]
	V	M,S	G,R		
Carteria klebsii, 1–27				CS	O
Family Chlamydomonidae					
Chlamydomonas sp., 1–27	×	×		CS	O
Family Volvacidae					
Pandorina morum, 1–27		×		CS	O
Order Euglenoidida					
Family Euglenidae					
Colacium sp., 1–27				CS	O
Trachelomonas sp., 1–27				CS	O
Phylum Annelida					
Class Enchytraeidae, 1, 2, 8					
Class Hirudinoidae, 2, 12, 25					
Class Lumbricidae, 2, 12, 25					
Family Lumbriculidae					
Class Oligochaeta					
Order Haplotaxida					
Family Naididae					
Nais pseudobtusa, 2	×	×	×	BSS	O
N. variabilis, 1, 2, 8	×	×	×	BSS	O
Nais sp., 2, 51.6 mi	×	×	×	BSS	O
Family Tubificidae					
Limnodrilus hoffmeisteri, 2, 12;		×		BSS	O
mile 51.6					
Tubifex tubifex, 1, 2; mile 51.6		×		BSS	O
Phylum Mollusca					
Order Basomatophora					
Class Gastropoda					
Family Lymnaeidae					
Fossaria obtrussa, 1, 2, 8; mile 51.6	×	×		SB	HD
Physella sp., 1–25	×	×	×	SB	C
Family Succineidae					
Oxyloma haydeni kanabensis, 2	×	×		SB	HD
Family Zonitidae					
Hawaiia miniscula, 2	×	×		SB	HD
Order Amphipoda					
Family Gammaridae					
Gammarus lacustris, 1–25	×	×		H	O
Class Insecta					
Order Diptera					
Family Ceratopogonidae					
Family Chironomidae					
Cricotopus annulator, 1–25	×	×	×	BSS	O
Cricotopus sp., 1, 2, 12, 20, 21, 22;	×	×	×	BSS	O
mile 51.6					
Eukiefferiella claripennis, 1–25	×	×	×	BSS	OC
E. coerulescens, 1–12	×	×	×	BSS	OC
Orthocladius luteipes, 2	×	×	×	BSS	HD
O. rivicola, 1–25	×	×	×	BSS	HD
Parakiefferiella sp., 2, 12, 20	×	×	×	BSS	O
Pseudosmitta sp., 2			×	BSS	O
Family Simuliidae					
Simulium arcticum, 1–25			×	BSS	O

TABLE 1.4. *Continued*

Taxon[a]	Substrate[b]			Source[c]	Feeding Habits[d]
	V	M,S	G,R		
Phylum Chordata					
Subphylum Vertebrata					
Class Osteichthyes					
Order Cypiniformes					
Family Catostomidae					
Catostomus latipinnis, widespread		×	×	CBT	O
Pantosteus (Catostomus) discobolus, widespread		×	×	CBT	O
Order Cypriniformes					
Family Cyprinidae					
Cyprinis carpio, widespread	×	×		CBT	O
Pimephales promelas, widespread		×		CBT	O
Rhinichthys osculus, widespread		×		CBT	O
Order Salmoniformes					
Family Salmonidae					
Salmo gairdneri, widespread			×	CBT	C
S. trutta, widespread			×	CBT	C
Salvelinus fontinalis, widespread			×	CBT	C

[a]An asterisk indicates an introduced species. Numbers indicate the sites where the samples were collected.
[b]V, vegetation and debris; M, mud; S, sand; G, gravel; R, rubble.
[c]BSS, Blinn et al. (1992); CB, Czarnecki and Blinn (1978); CBT, Czarnecki et al. (1976); CM, Carothers and Minckley (1981b); CS, Crayton and Sommerfeld (1978); H, Haury (1986); PP, Polhemus and Polhemus (1976); SB, Spamer and Bogan (1993).
[d]Feeding habits: C = carnivore; HD = herbivore-detritivore; O = omnivore; OC = omnivore-carnivore.

S. ulna var. *constricta*, which was present. Probably attached to sand or muddy particles were eight taxons belonging to the genus *Achnanthes*. These were widespread in this habitat (Table 1.4). Several *Cymbella* species were found associated with the mud and sand: *Cymbella prostrata* and *C. tumida*, which were common. Also present were *C. ventricosa* and *C. ventricosa* var. *semicircularis*. In quiet areas associated with the fine mud were the diatoms *Amphipleura pellucida*, *Anomoeoneis exilis*, *A. serians* var. *brachysira*, *A. vitrea*, and *Caloneis bacillum*. *Frustulia vulgaris* was also present associated with the mud and plant debris. Also in this habitat were *Navicula angelica*; *N. angelica* var. *subsalsa*, which was widespread; *N. arvenensis*; *N. cryptocephala*; *N. cryptocephala* forma. *minuta*; *N. cryptocephala* var. *veneta*; *N. decussis*; and *N. graciliodes*. All of these were very common in this habitat, that is, associated with fine sand and mud and organic debris. Other *Navicula* species common in this habitat were *N. mutica*; *N. mutica* var. *cohnii*; *N. pupula* var. *rectangularis*; *N. radiosa*, which was widespread; and *N. radiosa* var. *tenella*, also widespread. Other *Navicula* species were *N. subtilissima* and *Neidium dubium*. The family Nitzschiaceae, which is often associated with debris and mud, was represented by *Nitzschia acicularis*, *N. dissipata*, *N. frustulum*, and *N. romano*, and two species of *Surirella*, *Surirella angustata* and *S. brightwellei*.

Various protozoans were found associated with the fine mud and debris. They were the omnivores *Cryptomonas ovata*, *Dinobryon sertularia*, *Chlamydomonas* sp.,

and *Pandorina morum*. They consume plant material as well as very small protozoans and invertebrates.

Worms present in this habitat were *Nais pseudodobtusa*, *N. variabilis*, and an undescribed species of *Nais*. Tubificid worms in this habitat were the omnivorous
Limnodrilus hoffmeisteri and *Tubifex tubifex*. A few molluscs were found in this habitat: a herbivore–detritivore, *Fossaria obtrussa*; a carnivore, *Physella*; and the herbivore–detritivores *Oxyloma haydeni kanabensis* and *Hawaiia miniscula*. Also present
was an amphipod, *Gammarus lacustris* (Table 1.4).

Several Chironomidae were present in this habitat, which consisted of fine mud,
muddy sand, and organic debris, particularly plant debris. They were *Cricotopus annulator*, an omnivore; and an unidentified species of the same genus. The omnivore–
carnivores *Eukiefferiella claripennis* and *E. coerulescens* were common, as were the
herbivore–detritivores *Orthocladius luteipes* and *O. rivicola*. Another omnivore present was an unidentified species of *Parakiefferiella* sp.; most species of this genus
identified are omnivorous.

Several fish were found in this habitat, probably in pools associated with the fine
mud and sand. They were the omnivorous *Catostomus latipinnis*, which was widespread; and the omnivorous *Pantosteus* (*Catostomus discobolus*). Several cyprinids
were also found: *Cyprinus carpio*, which was widespread; *Pimephales promelas*,
an omnivore also introduced; and the widespread but probably native omnivore
Rhinichthys osculus.

Lotic Communities: Vegetation, Rocks, Gravel, and Coarse Sand

Mainly vegetation, but also rocks, gravel, and coarse sand formed excellent habitats
for many species in Vasey's Paradise. On the gravel, intermixed with mud, were
found *Oscillatoria articulata*, *O. limosa*, *O. nigra*, *Spirulina major*, *S. subsalsa*, and
Spirulina sp. These blue-green algae were fairly common and probably result from
use of this area for various human expeditions. Green algae, fairly common, were the
pollution-tolerant species of *Stigeoclonium flagelliferium* and an unidentified species
of *Stigeoclonium*. Also present was an unidentified species of *Cloniophora*. Present
were colonies of *Tetraedron* sp. Blue-green algae present in this habitat were *Microspora floccosa* and *Mougeotiopsis* sp.

The diatoms *Cyclotella meneghiniana* and *Melosira varians* were quite common.
Other species found in this habitat, which may have been growing there or perhaps
were deposited there, were *Asterionella formosa*, *Diatoma anceps*, *D. elongatum*,
D. hiemale, and *D. hiemale* var. *mesodon*, and common everywhere was *D. vulgare*.
Also in this habitat were found *Fragilaria construens*, *F. intermedia*, and *F. vaucheriae*, which was attached to the rocks or to gravel or debris in this habitat. Present
on the surface of the substrate, that is, gravel and sand, was *Synedra acus*, which was
very common. Also present were *S. nana*, *S. socia* (in the Little Colorado River, which
entered this habitat), and *S. tenera* var. *genuina*. *Synedra ulna* was widespread and
present was *S. ulna* var. *constricta*. Attached to the gravel and rocks in places where
the current was not very strong were *Achnanthes affinis*, *A. exigua* var. *heterovalva*,
A. lanceolata, *A. lanceolata* var. *dubia*, *A. linearis* forma. *curvata*, and *A. linearis*
var. *pusilla*. Also common in this habitat were *A. microcephala* and *A. minutissima*.

Attached to the rocks were also found *Cocconeis diminuta, C. pediculus, C. placentula* var. *euglypta,* and *C. placentula* var. *lineata.* Also common in this habitat was *Rhoicosphenia curvata.* Attached to the rocks and gravel were several taxons belonging to the family Cymbellaceae. They are food organisms for many types of aquatic invertebrates. The species present were *Amphora ovalis* var. *pediculus, Cymbella affinis,* and *C. amphicephala.* These three taxons were very common. Other members of the Cymbellaceae were found attached not only to rocks and gravel but also to filaments of algae or moss that lived in these habitats, where the flow was relatively rapid. They were, as seen in Table 1.4, 10 more taxons belonging to the genus *Cymbella.*

Also associated with the habitat where the flow was relatively rapid and rocks and gravel were the common substrates was *Frustulia vulgaris.* It probably occurred in areas between rocks or behind rocks where the flow was relatively slow. Widespread in this habitat were *Navicula cryptocephala, N. cryptocephala* forma. *minuta,* and *N. cryptocephala* var. *veneta.* Other common *Navicula* species were *Navicula decussis* and *N. minima.* Also common was *Navicula mutica.* Present in this habitat where the current was relatively swift was *N. mutica* var. *cohnii.* In among the rocks where the current was probably moderate were *N. tripunctata* and *Neidium dubium.* Also widespread in this habitat, where a current was present, were *Denticula elegans* and *D. rainerensis.* In among the rocks where the current was probably moderate were *Nitzschia amphibia,* which was widespread. Other common *Nitzschia* species were *N. dissipata, N. frustulum,* and *N. kutzingiana.*

Two species belonging to the family Ceratiaceae which are Dinophyceae were *Ceratium carolinianum* and *C. hirudinella.* Within the gravel were three worms believed to be omnivores. They belonged to the genus *Nais* and were *N. pseudobtusa, N. variabilis,* and an unidentified species of *Nais.* A single snail was found in this habitat, and it is believed to be a carnivore. It is an unidentified species of the genus *Physella.* In this habitat where the water was moving fairly rapidly but algae and other debris were present in among the rocks were found a number of chironomids. They were the omnivorous *Cricotopus annulator* and an unidentified species of the same genus. Also present were the omnivore–carnivores, *Eukiefferiella claripennis* and *E. coerulescens.* Two herbivore–detritivore taxons were also present: *Orthocladius luteipes* and *O. rivicola.* Other omnivores present were an unidentified species of *Parakiefferiella, Pseudosmitta* (an unidentified species), and the omnivorous *Simulium arcticum.*

Several species of fish were found in areas where the flow was fairly rapid: the omnivorous *Catostomus latipinnis* and *Pantosteus (Catostomus) discobolus.* They were widespread wherever the current was fairly rapid. A few species belonging to the family Salmonidae were there: *Salmo gairdneri, S. trutta,* and *Salvelinus fontinalis.* They are all carnivores believed to be introduced species.

SUMMARY

The Colorado River, which is 1360 miles (2190 km) in length, begins its long journey to the Gulf of California in Grand County, Colorado on the Continental Divide. It is formed by many small streams that receive their water from snow melting in the high Rocky Mountains. Downstream the Colorado is joined by the Gunnison, which is formed by the North and South Branches. The Uncompahgre River then joins the Colorado. Farther downstream the San Miguel and Dolores Rivers join the Colorado in northern Utah. The Green River, which is the longest tributary of the Colorado, joins it in Utah. The Green River rises in the Wind River Mountains of Wyoming. The Green River is joined by the Little Snake and the Yampa and White Rivers. Farther downstream the Green River is joined by the Duchesne, Price, San Rafael, and Dirty Devil Rivers. The Moab Canyon of Utah intersects with the Green River. Below the junction of the Green and Colorado Rivers is the Escalante River, which ends in the mouth of Lake Powell. The San Juan joins the Colorado in the region of Lake Powell. It is the largest river to join the Colorado. The San Juan is not only Colorado's largest river but also the largest river in New Mexico. The headwaters of the San Juan are in southwestern Colorado near Wolf Creek Pass. The San Juan receives not only many streams from Colorado but also streams that join it from New Mexico. Paria River joins the Colorado from 200 miles (321.87 km) south of the junction of the Colorado and the San Juan. Pipe Creek enters the Colorado River between river miles 88 and 89.

The vegetation along the Colorado River is sparse and consists of seep willow, Bermuda grass (*Cynodon dactylon*), and scratch grass. Hermit Creek enters the Colorado River from the south at river mile 95. Crystal Creek enters the river from the north rim between river miles 98 and 99. The Shinumo Creek enters the Colorado River from the northwest between river miles 108 and 109. The water temperature of Shinumo Creek at its mouth was 6 to 20°C when this study was made. However, the temperature of the water at the base of the waterfall was 10.5°C. Elves Chasm is located between river miles 116 and 117, where Royal Arch Creek enters the Colorado River.

Stone Creek enters the Colorado River from the north rim between river miles 131 and 132. It is an intermittent stream. Tapeats Creek enters the Colorado River near the north rim between river miles 133 and 134. Tapeats Creek contributes the largest discharge into the Colorado River from the north side of the Grand Canyon. Deer Creek enters the Colorado River from the north bank between river miles 136 and 137. Kanab Creek enters the Colorado River in Arizona between river miles 143 and 144. Kanab Creek is a sulfate stream low in nitrogen and phosphorus and relatively high in silica. 150 Mile Canyon, a small stream, flows intermittently into the Grand Canyon area between river miles 149 and 150. Near the mouth of the creek are small pools surrounded by seep willow, sawgrass, bullrush, and scratch grass. Above the falls of this creek, bullrushes, scratch grass, blue-eyed grass, and maidenhair fern line the canyon walls.

Havasu Creek enters the Colorado River from the south side of the canyon between river miles 156 and 157. This creek's water is dominated by calcium and magnesium carbonates and has a silicate concentration second only to that in Diamond Creek. Nitrogen and phosphorus values are low in Havasu Creek. Specific conductance is 660 to 740 μS cm^{-1}. The pH ranges from 8.3 to 9.0. National Canyon enters the Colorado River from the southeast between river miles 166 and 167.

Three Spring Canyon enters the Colorado River between river miles 215 and 216. It is an intermittent creek. The stream 217 Mile Canyon enters the canyon between river miles 217 and 218.

Diamond Creek enters the Colorado River from the north between river miles 225 and 226. At its mouth it is wide with little vegetation. Seep willow, cattails, and rabbit foot grass occur at or near the margins of the stream. The water temperature at the mouth ranges from 12° to 27°C. An intermittent stream, Bridge Canyon, enters the river between river miles 235 and 236.

Above Hoover Dam at Grand Wash (river mile 276), the river turns sharply right and flows south again, forming the boundary between Arizona and Nevada. In this area it is joined by the Virgin River from the Zion Canyon and the Bill Williams River, which drains 5400 m^2 in Arizona. Near Yuma, the Colorado receives the Gila River, which rises in the Elk Mountains. It is 1010 km long and one of the most important rivers in America's Southwest. Coolidge Dam was constructed on the Gila River in 1928. Its reservoir holds about 6000 million feet of water. The Gila is Arizona's largest river and drains approximately 56,000 square miles. Downstream from Hoover Dam is Davis Dam. The Colorado downstream forms the boundaries between California and Arizona. The next humanmade impediment is Parker Dam at the southeastern end of Lake Havasu. Parker Dam is the beginning of the Colorado Aqueduct, which carries the water to the city of Los Angeles and other areas of southern California. Besides the large reservoirs mentioned above, the Imperial and Leguna Reservoirs, which are largely for irrigation, are formed on the Colorado River. In Mexico, the Colorado flows between the area known as the Sonora and the mainland from the upper district of the penninsula of lower California. Hence it moves slowly into the Gulf of California and sinks its red waters into the blue waters of the ocean.

The Colorado River has three distinct sections: the Upper Basin, the plateau section, and the Lower Basin. Many dams have been built on the Colorado River system. The major dams on the Colorado proper are the Glen Canyon Dam and Hoover Dam. Smaller dams are also located on the Colorado River. The geology of the Colorado River basin is highly varied and is composed of igneous, metamorphic, and sedimentary rock types as well as clays. The soils of the basin resemble the geological formations from which they are derived. The geology of the Grand Canyon is highly varied in the types of rocks and sands that it contains.

The biology of the watershed of the Colorado River is best known between the Glen Canyon and Hoover Dams. Woody species of the riparian area contribute greatly to the support of the insects and other invertebrates present because this area is less subject to floods and droughts. The present native species are a minor component of the overall flora because of introduced species. There is an ever-increasing predominance of salt cedar, camelthorn, and herbaceous species such as yellow clover and spiny sow thistle. A nonnative plant along the river is tamarisk, commonly known as salt cedar. The salt cedar is a poor habitat for most wildlife except for the mourning dove and the white-winged dove. The salt cedar attracts insects as well as riparian birds. Another plant of import is the camelthorn, a spiny leguminous half-shrub that has invaded several beaches, especially in the upper portion of the canyon. The Russian olive is also fairly common in the canyon. It attracts the mourning dove as well as other bird species. Other species important to wildlife are the western honey mosquite, the catclaw acacia, and the coyote or sandbar willow. Seep willows are represented by two species: *Baccharis saliscifolia* is common along the river and *B. emoryii* along the sidestreams. The herbaceous riparian species are foxtail broom and Bermuda grass, as well as several other plants that have been discovered in this reach: that is, the Grand Canyon reach.

The amphibians in the Colorado River are usually associated with moist habitats: one species of salamander and five species of toads and frogs, four of which were found. Grand Canyon rattlesnake or pink rattlesnake occurs frequently in riparian habitats of the Colorado River. Seven species of lizards are found in the riparian zone. Because of greater vegetation biomass, the reptile population has been increased since the river flow has been stabilized. An increase in insects has also occurred. In 1978, 284 species of birds were known from the Grand Canyon. Most of the species of birds that are found are insectivorous, and therefore the relatively high insect populations and low human populations probably account for their being present. Seventy-eight species of birds have been reported by Hoffmeister (1971). Forty of these occurred in the inner gorge, and most of these occurred in the riparian zone. Two endangered bird species of the riparian zone are the bald eagle and the peregrine falcon. The mammals are restricted to riparian mammals such as beavers, or to a lesser amount to bighorn sheep.

Dam-related effects such as clarity, pattern of flow, and water temperature override the geomorphological influences on habitat availability. The recovery of the benthos did not seem to be correlated with the abundance of the fish but rather with the geomorphological differences in substrate availability between reaches, mediated by dam and tributary effects on water clarity and the amount of the benthos. Reaction between flow regulation and geomorphology produced a pattern of circuitous recovery of some physical river systems characteristic of distance from the dam. Improving discharge management for endangered native fish species requires detailed understanding of the existing and potential benthic development and trophic interactions. Many factors are involved in recovery of the benthos and hence the fish populations.

Vegetation is necessary for mammals, and mammals have a profound effect on the existing vegetation. For example, severe damage to vegetation may be created by foraging and trampling. This is particularly true of feral animals. The lack of establishment of Goodding willows and Fremont cottonwoods is probably due to beavers. Seedling establishment is not well understood for most woody riparian species. Grand Canyon species are short on endemics and rare species. Two previously undescribed species of flowering plants, *Flaveria mcdougalii* and *Euphorbia aaron-rossii*, are such plants. The riparian vegetation of the Grand Canyon is very important to the aquatic life of the river. It is the high diversity of the aquatic life that supports the high diversity of birds and amphibia, and the aquatic insects are very important to the fish.

Associations of Aquatic Organisms: Glen Canyon Dam to Hoover Dam.   This reach of the river is often known as the Grand Canyon reach. The algae are sometimes free-floating and seem to be part of the attached vegetation, which has been broken off by flow. The algae may be found on fine sediments in backwaters within submerged vegetation. The areas of algae attachment were the scoured rock faces in areas of rapids and cataracts and fine sediments in backwaters, usually along the inner side of river bends. Various submerged macrophytes also serve as habitats for algae such as diatoms. Many of the plankton species are believed to be detached from various surfaces. The biodensity of diatom species of the Colorado River after impoundment was 1600-fold lower than the cell density prior to impoundment of the river.

Crayton and Sommerfeld (1978) reported 127 species of phytoplankton in the Colorado River. Many of these were detached and dislodged from substrates. The dominant species of diatoms were *Diatoma vulgare*, *Rhoicosphenia curvata*, and *Cocconeis pediculus*. Many of these algae are known to have been detached forms from substrates in Lake Powell.

Cladophora glomerata is the dominant attached filamentous green algae in the canyon, especially between Glen Canyon Dam and Paria River and at the mouths of tributaries. This algae was also the dominant algae for sites at and above Lees Ferry. A relatively high amount of biomass of *Cladophora glomerata* in upstream tailwater sites may be the result of stable rock faces for attachment and nutrient-enriched waters. Abrupt drops in *Cladophora* at Lees Ferry were probably due to episodes of desiccation and reduced input of nutrients. *Cladophora* seems to thrive under continuously submerged clear-water habitats. *Cladophora glomerata* seems to prefer shallow waters and decreases with greater depth, probably due to rapid attenuation of light caused by periodically high sediment loads. There seemed to be some correlation in *Cladophora glomerata* with channel depth at sites above Lees Ferry. Similarly, there was a decrease in biomass with depth at sites below Lees Ferry.

Cladophora glomerata was found to be a habitat for the amphipod *Gammarus lacustris* and other snails and invertebrates. It also provides a large amount of surface for the attachment of epiphytic diatoms, which are an important food source for aquatic invertebrates and, in some cases, for fish.

The diatoms were the dominant form of algae, followed by cyanobacteria and chlorophytes. At the confluence of Diamond Creek, a red alga, *Batrachospermum* sp., was reported, and *Audouienella* sp. was found attached to filaments of *Cladophora*. The dominant diatom taxa were *Diatoma vulgare*, *Synedra ulna*, and *Cocconeis pediculus*. *Achnanthes affinis*, *Cocconeis pediculus*, *Diatoma vulgare*, and *Rhoicosphenia curvata* made up 80% of the communities upstream from Lees Ferry. However, these four taxa were less important at downstream sites, and *Gomphonema olivaceum*, *Cymbella affinis*, and *Nitzschia dissipata* became more important at the downstream sites. It is believed that the change in species composition was due to their tolerance to suspended sediments.

Below Lees Ferry there was a fourfold decrease in epiphytic diatoms. It is difficult to compare the aquatic invertebrates before and after dam closure because of the introduction of many individuals of insects, snails, and leeches. The zooplankton and the algae seemed to be derived from lentic populations in Lake Powell. There seemed to be some correlation with the increase in populations of *Cladocera* and copepods with the temporary rise in water due to patterns of releases from the dam. Below the dam, the microcrustacea also increased. A large percent of the copepod planktors were in poor condition and increase downstream.

Of the 34 invertebrates listed by Haury (1986) only 16 were true planktors. Others were the benthic species. These species, which have broken loose and drifted downstream, contributed most of the biomass. Other planktors listed by Kubly are rotifers (*Collembolan*) and water mites. The last two are characterized best as epineustonic. The mites are probably nectonic.

Backwaters and borders of the rivers appeared to be refugia for some invertebrates. There seemed to be a higher productivity of these invertebrates in the backwaters. Most of the sites for aquatic life are located near the mouths of tributaries and around debris caught in the river in the Grand Canyon National Park and vicinity. Just below Glen Canyon Dam, the current is stochastic and the water is generally clear and deep. The rapid flow inhibits many different types of organisms.

The habitats for algae consisted mainly of other algae, submerged leaves, and stems of vascular plants. Some mosses also were habitats of algae. A green alga, *Cladophora glomerata*, is much more prevalent than it was in the early days of the Colorado River and seems to have been increased because of the dams, resulting in clearer water with increased light penetration and greater stability of the riverbed.

Seasonal extremes in discharged sediment loads and temperature seemed to have been eliminated since the construction of Grand Canyon Dam. The discharge is now regulated by power demands. The maximum is about 566 cm s^{-1} and the daily minimum is 130 cm/s^{-1}, with extremes ranging from 28 to 764 cm/s^{-1}. Water entering the river below the dam is cold and clear. The temperature ranges from 15 to 60°C.

Invertebrates in the main stem of the Colorado River are generally low in productivity, except in the section between Glen Canyon Dam and the confluence of the Little Colorado River, where the sediment input was minimum. Here the estimates of density were several thousand individuals per square meter. In contrast, grabs taken below the confluence of the Little Colorado River usually yielded only 5 to 10 individuals per square meter. These were blackflies, midges, and aquatic earthworms, which were collected on exposed gravel in the stream channel and along the margins of the river. The density of invertebrates seemed to be lower at the confluence of tributaries. Insects are fewer in spring and summer. The productivity in the main stem of the Colorado River seemed to be lower than in the tributaries and other riverine systems. This seems to be caused by the cooler annual temperature, the greater depth, the current velocity, and sediment input. Common invertebrates in the main stream were freshwater amphipods and aquatic earthworms.

The diversity of insects seemed to be lower than expected, probably because high canyon walls to the north and south, combined with the topographic variability created a formidable dispersal barrier for aquatic species. Canyon topography has effectively blocked northern dispersal of some species. Spring and summer productivity and diversity were lowest. It is probably due to spring runoff and summer flash floods from the tributaries that disrupt the benthic invertebrate communities. The small amount of flooding in fall and winter favored algae and invertebrate productivity. The main stream of the Colorado River is typically cooler than the tributaries.

This may have greatly influenced the invertebrate fauna. Furthermore, turbidity and suspended solids were higher in the main stream than in the tributaries. It is undoubtedly a combination of unstable physical and chemical parameters that reduces the fauna in the mouths of the tributaries and in the Colorado River just below the entrances of tributaries. In general, the productivity of invertebrates was lower in the main channel than in the tributaries. This is due to the river's cooler annual temperature, greater depth, current velocity, and sediment input.

The areas that produce the richest invertebrate fauna are shallow backwaters and eddies along the margin of the river. The influence of physical and chemical effects produced by the diurnal water fluctuations of the Colorado River on the invertebrate fauna at the confluences appears to be substantial. The instability of the substrates also contributes to reducing the invertebrate fauna in diversity and productivity.

The fish fauna of the Colorado River has been studied primarily in the region of the Grand Canyon and in Lake Powell and Lake Mead. The Colorado River Basin contains a mixture of native and introduced species. Over 78% of the species of fish now known from the basin are particular to it. This is a larger percentage of species particular to a given river than that found in any other river in the United States. This is probably because of its long-term isolation and its unusual physical characteristics. There are several distinctive features about the fish fauna of the Colorado River. The larger fish seem to live longer than in other rivers, and they are more streamlined and fusiform than most fish. Many have small depressed skulls, large predorsal humps or keels or both, and elongated pencil-thin caudal peduncles. Typically, they have small eyes, expansive and falcated fins, and thick leathery skin. Scales are thin and deeply embedded in the skin, and sometimes they almost seem to be absent. These special characteristics seem to be adapted to the severe habitats of the Colorado River. These characteristics seem to be important in allowing the fish to maintain position and maneuver itself in swift turbulent currents. The small eyes help to reduce the erosive forces on the eye.

In 1960, the native fish populations of the lower Colorado had been largely replaced by exotic species due to changes of habitat and general characteristics of the water. The humpback chub, bonytail, Colorado spikeminnow, and razorback suckers are listed or proposed as endangered by the Department of Interior (U.S. Fish and Wildlife Service, 1988). Of these species only the humpback chub has reproductive populations. Other fish are extremely extirpated. Speckled dace, flannelmouth suckers, and blue suckers remain relatively common. The fish fauna is best known between Lees Ferry and Separation Rapids. Twenty-seven fish species are known from the study area, of which 70% are exotics. Originally in the Colorado River there were eight native species of fish, including the bonytail chub, Colorado roundtailed chub, and Colorado spikeminnow, which are apparently extinct in the Grand Canyon. Humpback chub is also endangered. The razorback sucker is thought to be going extinct.

Although 19 exotic species have been introduced into the Colorado River, 10 of the 19 have been observed so infrequently that they should be considered as an insignificant component of the Grand Canyon ichthyofauna. Exotic minnow species that seem to be well established in the Grand Canyon are the fathead minnow and the

Rio Grande killifish. Species of trout have been introduced by the sports fisheries stocking programs. Rainbow trout is most widely distributed in the area. The brown and brook trout are far less numerous. Cutthroat trout were introduced at Lees Ferry. In November 1979 it was the most common fish taken by anglers at Badger Rapids.

Channel catfish are probably ubiquitous throughout the study area, although they are hard to catch. Twenty-seven species of fish are known to be present or have occurred in the Colorado River and its tributaries in the Grand Canyon reach.

The fish fauna is well sorted as to its feeding habits. Some feed largely on algae, whereas others feed on a variety of invertebrates, insects, and even small fish. Besides aquatic insects, the fish fauna feed upon terrestrial insects. The most heavily exploited are stinkbugs, grasshoppers, ants, and scarab beetles. Thirty-six types of food have been identified in the stomach of the rainbow trout taken in the southern tributaries. The food was strikingly different in these tributaries from that of the main stream and northern tributaries. *Cladophora* was present in small quantities during fall and winter and absent from the diet in spring and summer. Mayflies were the major prey. Pouch snails were the most common aquatic invertebrate present in the stomach contents during the summer. Common Diptera in the diet included blackfly, soldier fly, midge, horsefly, and moth fly larvae, as well as midge and blackfly pupae. Other aquatic invertebrates consumed were diving beetles, sea shrimp, net-spinning caddis, and scuds. Of the terrestrial insects, grasshoppers, blackflies, and bees were common prey, particularly in the summer and fall. Most of the fish fed on a variety of foods, particularly insects. However, the carp feed primarily on *Cladophora* during all seasons. Scuds and midge larvae are the most commonly ingested aquatic invertebrates, but they form less than 12% of the seasonal diet. The fish fauna consists of a variety of species feeding on different forms of aquatic life, particularly insects, and in some cases, a fair amount of algae was consumed.

The structure of the aquatic communities is described for Vasey's Paradise ecosystem as an example. The basic structure of the ecosystems in these various habitats is based on algae. Protozoans and small invertebrates are detritivores, herbivores or omnivores. Larger invertebrates and a few fish are omnivores or carnivores. In other words, the ecosystem consisting of four stages of nutrient and energy transfer is well exemplified in Vasey's Paradise, which illustrates the general ecosystems in the Grand Canyon reach.

BIBLIOGRAPHY

Adams, D., and V. Lamarra, eds. 1983. Aquatic resource management of the Colorado River ecosystems. Ann Arbor Science, Ann Arbor, Mich.

Alstead, D. N. 1980. Comparative biology of the common Utah Hydropsychidae (Trichoptera). Am. Midl. Nat. 103(1): 167–174.

Ayers, A. D., and T. McKinney. 1996. Water chemistry and zooplankton in the Lake Powell forebay, Glen Canyon Dam discharge and tailwater. Final Report. Arizona Game and Fish Department. Phoenix, Ariz. 70 pp. Unpublished.

Bahls, L. L., E. E. Weber, and J. O. Jarvie. 1984. Ecology and distribution of major diatom ecotypes in the southern Fort Union Coal Region of Montana. U.S. Geological Survey Professional Paper 1289: 1–151.

Baldys, S., L. K. Ham, and K. D. Fossum. 1995. Summary statistics and trend analysis of water-quality data of sites in the Gila River basin, New Mexico and Arizona. U.S. Geological Survey Water-Resources Investigations Report 95-4083.

Bateman, A. M. 1938. The Colorado delta. Econ. Geol. 33: 119–120 [book review].

Behnke, R. J., and D. E. Benson. 1980. Endangered and threatened fishes of the upper Colorado River basin. Colorado State Univ. Cooperative Extension Service Bulletin 503A.

Benenati, P. L., J. P. Shannon, and D. W. Blinn. 1998. Desiccation and recolonization of phytobenthos in a regulated desert river: Colorado River at Lees Ferry, Arizona, U.S.A. Regul. Rivers Res. Manag. 14: 519–532.

Blinn, D. W. 1984. Growth responses to variations in temperature and specific conductance by *Chaetoceros muelleri* (Bacillariophyceae). Br. Phycol. J. 19: 31–35.

Blinn, D. W. 1993. Diatom community structure along physicochemical gradients in saline lakes. Ecology 74(4): 1246–1263.

Blinn, D. W., and G. A. Cole. 1991. Algal and invertebrate biota in the Colorado River: comparison of pre- and post-dam conditions. *In* Committee to Review the Glen Canyon Environmental Studies, Colorado river ecology and dam management: proceedings of a symposium, May 24–25, 1990, Santa Fe, N. Mex. National Academy Press, Washington, D.C.

Blinn, D. W., R. E. Truitt, and A. Pickart. 1989a. Feeding ecology and radular morphology of the freshwater limpet, *Ferrissia fragilis*. J. North Am. Benthol. Soc. 8(3): 237–242.

Blinn, D. W., R. Truitt, and A. Pickart. 1989b. Short communications. Response of epiphytic diatom communities from the tailwaters of Glen Canyon Dam, Arizona to elevated water temperature. Regul. Rivers Res. Manag.. 4: 91–96.

Blinn, D. W., L. E. Stevens, and J. P. Shannon. 1992. The effects of Glen Canyon Dam on the aquatic food base in the Colorado River corridor in Grand Canyon, Arizona. Glen Canyon Environmental Study-II-02. 100 pp.

Blinn, D. W., R. H. Hevly, and O. K. Davis. 1994a. Continuous holocene record of diatom stratigraphy, paleohydrology, and anthropogenic activity in a spring-mound in southwestern United States. Quat. Res. 42: 197–205.

Blinn, D. W., J. P. Shannon, K. Wilson, and L. E. Stevens. 1995. Interim flow effects from Glen Canyon Dam on the aquatic food base in the Colorado River in Grand Canyon National Park, Arizona. In cooperation with the Glen Canyon Environmental Studies Program, National Park Service Cooperative Agreement CA 8024-8-0002.

Blinn, D. W., J. P. Shannon, L. E. Stevens, and J. P. Carder. 1995a. Consequences of fluctuating discharge for lotic communities. J. North Am. Benthol. Soc. 14(2): 233–248.

Blinn, D. W., F. Govedich, and S. Earl. 1996. Preliminary report on an aquatic ecosystem assessment of four major tributaries in the Gila Wilderness Area, New Mexico: role of fire on stream ecosystems. Northern Arizona Univ. Press, Flagstaff, Ariz.

Brown, R. M. 1927. The utilization of the Colorado River. Geogr. Rev. 17: 453–466.

Brown, R. M. 1932. Complications in the utilization of the Colorado River. Geogr. Rev. 22: 317.

Brown, E. H., and R. S. Babcock. 1974. A preliminary report on the geology of the older Precambrian rocks in the upper granite gorge of the Grand Canyon. Proceedings of the Geological Society of America Rocky Mountain Section 27th annual meeting, Flagstaff, Ariz. 6(5): 429–430.

Brown, B. T., and L. E. Stevens. 1997. Winter bald eagle distribution is inversely correlated with human activity along the Colorado River, Arizona. J. Raptor Res. 31(1): 7–10.

Brown, B. T., S. W. Carothers, and R. R. Johnson. 1987. Grand Canyon birds: historical notes, natural history, and ecology. Univ. Arizona Press. Tucson, Ariz.

Carothers, S. W., and R. R. Johnson. 1975. Recent observations on the status and distribution of some birds of the Grand Canyon region. Plateau 47: 140–153.

Carothers, S. W., and C. O. Minckley. 1981a. A survey of the aquatic flora and fauna of the Grand Canyon. Final Report. U.S. Department of the Interior, Water and Power Resources Service, Lower Colorado Region, Boulder City, Nev.

Carothers, S. W., and C. O. Minckley. 1981b. A survey of the fishes, aquatic invertebrates, and aquatic plants of the Colorado River and selected tributaries from Lees Ferry to Separation Rapids. Final Report.

U.S. Department of the Interior, Water and Power Resources Service, Lower Colorado Region, Boulder City, Nev. 401 pp.

Carothers, S. W., S. W. Aitchison, and R. R. Johnson. 1976. Natural resources, white water recreation, and river management alternatives of the Colorado River, Grand Canyon National Park. First conference on scientific research in the national parks, New Orleans, La.

Carter, J. C., R. A. Valdez, R. J. Ryel, and V. A. Lamarra. 1985. Fisheries habitat dynamics on the upper Colorado River. J. Freshw. Ecol. 3(2): 249–264.

Carter, J. C., V. A. Lamarra, and R. J. Ryel. 1986. Drift of larval fishes in the upper Colorado River. J. Freshw. Ecol. 3(4): 567–577.

Cole, G. A., and D. M. Kubly. 1976. Limnologic studies on the Colorado River from Lees Ferry to Diamond Creek. Colorado River Research Series Contribution 37. 99 pp.

Colletti, P. J., D. W. Blinn, A. Pickart, and V. T. Wagner. 1987. Influence of different densities of the mayfly grazer *Heptagenia criddlei* on lotic diatom communities. J. North Am. Benthol. Soc. 6(4): 270–280.

Collins, W. D., and C. S. Howard. 1927. Quality of water of Colorado River in 1925–1927. U.S. Geological Survey Water Supply Paper 596-B: 33–43.

Colorado River Wildlife Council. 1977. Endemic fishes of the Colorado River system: a status report. CRWC, Colo. 16 pp.

Committee to Review the Glen Canyon Environmental Studies. 1991. Colorado River ecology and dam management: proceedings of a symposium, May 24–25, 1990, Santa Fe, N. Mex. National Academy Press, Washington, D.C.

Committee to Review the Glen Canyon Environmental Studies. 1996. River resource management in the Grand Canyon. National Resource Council. National Academy Press, Washington, D.C.

Crayton, W. M., and M. R. Sommerfeld. 1978. Phytoplankton of the lower Colorado River, Grand Canyon region, J. Ariz.–Nev. Acad. Sci. 13(1): 19–24.

Czarnecki, D. B., and D. W. Blinn. 1978. Diatoms of the Colorado River in the Grand Canyon National Park and vicinity. Bibl. Phycol. 38. 181 pp.

Czarnecki, D. B., D. W. Blinn, and T. Tompkins. 1976. A periphytic microflora analysis of the Colorado River and major tributaries in Grand Canyon National Park and vicinity. Colorado River Research Program Final Report. Technical Report 6. Grand Canyon National Park, National Park Service, U.S. Department of the Interior.

Dolan, R., A. Howard, and A. Gallenson. 1974. Man's impact on the Colorado River in the Grand Canyon. Am. Sci. 62: 392–401.

Dolan, R., A. Howard, and D. Trimble. 1978. Structural control of the rapids and pools of the Colorado River in the Grand Canyon. Science 202: 629–631.

Duncan, S. W., and D. W. Blinn. 1989. Importance of physical variables on the seasonal dynamics of epilithic algae in a highly shaded canyon stream. J. Phycol. 25: 455–461.

Fortier, S., and H. F. Blaney. 1928. Silt in the Colorado River and its relation to irrigation. U.S. Department of Agriculture Techical Bulletin 67. 97 pp.

Fuller, R. L., and K. W. Stewart. 1979. Stonefly (Plecoptera) food habits and prey preferences in the Dolores River, Colorado. Am. Midl. Nat. 101(1): 170–181.

Gehlbach, F. R. 1966. Grand Canyon amphibians and reptiles: field check list. Grand Canyon Natural History Association, Grand Canyon, Ariz. 4 pp.

Ghassemi, F., A. J. Jakeman, and H. A. Nix. 1995. Salinisation of land and water resources: human causes, extent, management, and case studies. CAB International, Wallingford, Berkshire, England.

Grater, R. 1937. Checklist of birds of Grand Canyon National Park. Grand Canyon Natural History Association Bulletin 8. 55 pp.

Hamblin, W. K., and J. K. Rigby. 1968. Guidebook to the Colorado River, Part 1: Lees Ferry to Phantom Ranch in Grand Canyon National Park. Brigham Young Univ. Geol. Stud. 15(5): 1–84.

Harbeck, G. E., M. A. Kohler, G. E. Koberg, et al. 1958. Water-loss investigations: Lake Mead studies. U.S. Geological Survey Professional Paper 298. 100 pp.

Hardwick, G. G., D. W. Blinn, and H. D. Usher. 1992. Epiphytic diatoms on *Cladophora glomerata* in the Colorado River, Arizona: longitudinal and vertical distribution in a regulated river. Southwest. Nat. 37(2): 148–156.

Hart, B. T., P. Bailey, R. Edwards, K. Hortle, K. James, A. McMahon, C. Meredith, and K. Swadling. 1990. Effects of salinity on river, stream, and wetland ecosystems in Victoria, Australia. Water Res. 24: 1103–1117.

Haury, L. R. 1986. Zooplankton of the Colorado River: Glen Canyon Dam to Diamond Creek. Report B-10. Aquatic Biology of the Glen Canyon Environmental Studies. 58 pp.

Hoffmeister, D. F. 1971. Mammals of Grand Canyon. Univ. Illinois Press, Urbana, Ill. 183 pp.

Holden, P. B. 1968. Systematic studies of the genus *Gila* (Cyprinidae) of the Colorado River basin. Unpublished master's thesis. Utah State Univ.–Logan. 68 pp.

Holden, P. B. 1979. Ecology of riverine fishes in regulated stream systems with emphasis on the Colorado River. pp. 57–74. *In* J. V. Ward and J. A. Stanford, eds., The ecology of regulated streams. Plenum Press, New York. 398 pp.

Holden, P. B., and C. B. Stalnaker. 1975. Distribution and abundance of mainstream fishes of the middle and upper Colorado River basins, 1967–1973. Trans. Am. Fish. Soc. 104(2): 217–231.

Howard, C. S. 1929. Suspended matter in the Colorado River in 1925–1928. U.S. Geological Survey Water Supply Paper 636-B: 15–44.

Howard, C. S. 1931. Quality of water of the Colorado River in 1928–1930. U.S. Geological Survey Water Supply Paper 638-D: 145–158.

Iorns, W. V., C. H. Hembree, and G. L. Oakland. 1965. Water resources of the upper Colorado River basin. Technical report. U.S. Geological Survey Professional Paper 441.

Johnson, R. R. 1991. *In* Colorado River ecology and dam management: proceedings of a symposium, May 24–25, 1990, Santa Fe, N. Mex. National Academy Press, Washington, D.C. pp. 124–177.

Jordan, D. S. 1891. Report of explorations in Colorado and Utah during the summer of 1889, with an account of the fishes found in each of the river basins examined. Bull. U.S. Fish. Comm. 9: 1–40.

Joseph, T., J. Sinning, R. Behnke, and P. Holden. 1977. An evaluation of the status, life history, and habitat requirements of endangered and threatened fishes of the upper Colorado River system. Fish and Wildlife Service/Office of Biological Services Report 24.

Korte, V. L., and D. W. Blinn. 1983. Diatom colonization on artificial substrata in pool and riffle zones studied by light and scanning electron microscopy. J. Phycol. 19: 332–341.

Kubly, D. M. 1976. Personal communication to D. W. Blinn.

Lamarra, V. A., M. C. Lamarra, and J. C. Carter. 1985. Ecological investigation of a suspected spawning site of Colorado squawfish on the Yampa River, Utah. Great Basin Nat. 45(1): 127–140.

Leibfried, W. C., and D. W. Blinn. 1986. The effects of steady versus fluctuating flows on aquatic macroinvertebrates in the Colorado River below Glen Canyon Dam, Arizona. NTIS Report PB88206362/AS.

Leopold, L. B. 1969. The rapids and the pools—Grand Canyon. *In* The Colorado River region and John Wesley Powell. U.S. Geological Survey Professional Paper 669-D: 131–145.

Lowe, B. 1974. Environmental requirements and pollution tolerance of freshwater diatoms. U.S. Environmental Protection Agency, Washington, D.C. 334 pp.

MacCracken, J. G. 1981. Notes: the diatom flora of the lower Chevelon Creek area of Arizona: an inland brackish water system. Southwest. Nat. 26(3): 311–324.

McAda, C. W., and R. S. Wydoski. 1985. Growth and reproduction of the flannelmouth sucker, *Catostomus lattipinnis*, in the upper Colorado River basin, 1975–76. Great Basin Nat. 45(2): 281–286.

Miller, R. R. 1952. Bait fishes of the lower Colorado River, from Lake Mead, Nevada, to Yuma, Arizona, with a key for their identification. Calif. Dep. Fish Game Bull. 38: 7–42.

Minckley, W. L. 1985. Native fishes and natural aquatic habitats in U.S. Fish and Wildlife Service Region II, west of the Continental Divide. Internal Report. U.S. Fish and Wildlife Service, Division of Endangered Species, Albuquerque, N. Mex.

Minckley, W. L. 1991a. Native fishes of arid lands: a dwindling resource of the desert Southwest. U.S. Department of Agriculture, Forest Service, Rocky Mountain Forest and Range Experiment Station, Fort Collins, Colorado. 45 pp.

Minckley, W. L. 1991b. Native fishes of the Grand Canyon region: an obituary? *In* Colorado river ecology and dam management: proceedings of a symposium, May 24–25, 1990, Santa Fe, N. Mex. National Academy Press, Washington, D.C. pp. 124–177.

Minckley, T. A., O. K. Davis, C. Eastoe, and D. W. Blinn. 1997. Analysis of environmental indicators from a mastodon site in the Prescott National Forest, Yavapai County, Arizona. J. Ariz.–Nev. Acad. Sci. 30(1): 23–29.

National Research Council. 1996. River resource management in the Grand Canyon. National Academy Press, Washington, D.C.

New Mexico Water Quality Control Commission. 1974. Lower Colorado River basin plan. The Commission, Santa Fe, N. Mex.

Oberlin, G. E., J. P. Shannon, and D. W. Blinn. 1999. Watershed influence on the macroinvertebrate fauna of ten major tributaries of the Colorado River through Grand Canyon, Arizona. Southwest. Nat. 44: 17–30.

Ohmart, R. D., B. W. Anderson, and W. C. Hunter. 1988. The ecology of the lower Colorado River from Davis Dam to the Mexico–United States International Boundary: a community profile. U.S. Fish and Wildlife Service Biological Report 85(7.19). 296 pp.

Polhemus, J. T., and M. S. Polhemus. 1976. Aquatic and semi-aquatic Heteroptera of the Grand Canyon (Insect: Heteroptera). Great Basin Nat. 36: 221–226.

Propst, D. L. 1982. Warmwater fishes of the Platte River basin, Colorado: distribution, ecology, and community dynamics. Doctoral dissertation, Colorado State Univ. 285 pp.

Propst, D. L., and C. A. Carlson. 1986. The distribution and status of warmwater fishes in the Platte River drainage, Colorado. Southwest. Nat. 31(2): 149–167.

Propst, D. L., and C. A. Carlson. 1989. Life history notes and distribution of the Johnny darter, *Etheostoma nigrum* (Percidae), in Colorado. Southwest. Nat. 34(2): 250–259.

Rabbitt, M. C., E. D. McKee, C. B. Hunt, and L. B. Leopold. 1969. The Colorado River region and John Wesley Powell. U.S. Geological Survey Professional Paper 669.

Reisner, M. 1993. Cadillac desert: The American West and its disappearing water. Penguin Books, New York.

Rinne, J. N., and W. L. Minckley. 1991. Native fishes of arid lands: a dwindling resource of the desert Southwest. U.S. Department of Agriculture Forest Service General Techical Report RM-206.

Robinson, A. T., D. M. Kubly, R. W. Clarkson, and E. D. Creef. 1996. Factors limiting the distributions of native fishes in the Little Colorado River, Grand Canyon, Arizona. Southwest. Nat. 41(4): 378–387.

Shannon, J. P., D. W. Blinn, and L. E. Stevens. 1994. Trophic interactions and benthic animal community structure in the Colorado River, Arizona. Freshw. Biol. 31: 213–220.

Shannon, J. P., D. W. Blinn, K. P. Wilson, P. L. Benenati, and G. Oberlin. 1996. Interim flow and beach building spike flow effects from Glen Canyon Dam on the aquatic food base in the Colorado River in Grand Canyon National Park, Arizona. In cooperation with the Glen Canyon Environmental Studies Program. 111 pp.

Shaver, M. L., J. P. Shannon, K. P. Wilson, P. L. Benenati, and D. W. Blinn. 1997. Effects of suspended sediment and desiccation on the benthic tailwater community in the Colorado River, U.S.A. Hydrobiologia 357: 62–72.

Short, R. A. 1979. Particulate organic matter dynamics in a Colorado mountain stream. Ph.D. dissertation. Colorado State Univ.

Short, R. A., S. P. Canton, and J. V. Ward. 1980. Detrital processing and associated macroinvertebrates in a Colorado mountain stream. Ecology 61(4): 727–732.

Sommerfeld, M. R., W. M. Crayton, and N. L. Crane. 1976. Survey of bacteria, phytoplankton, and trace chemistry of the Lower Colorado River and tributaries in the Grand Canyon National Park. Technical Report 12. Colorado River Research Program. Report Series. Grand Canyon National Park.

Soule, J. M. 1992. Precambrian to earliest Mississippian stratigraphy, geologic history, and paleography of northwestern Colorado and west-central Colorado. U.S. Geological Survey Bulletin 1787-U.

Spamer, E. E. 1981. Bibliography of the Grand Canyon and lower Colorado River: 1540–1980. Grand Canyon Natural History Association, Grand Canyon, Ariz. 119 pp.

Spamer, E. E., and A. E. Bogan. 1993. Mollusca of the Grand Canyon and vicinity, Arizona: new and revised data on diversity and distributions, with notes on Pleistocene–Holocene mollusks of the Grand Canyon. Proc. Acad. Nat. Sci. Phila. 144: 21–68.

Stanford, J. A., and J. V. Ward. 1983. The effects of regulation on the limnology of the Gunnison River: a North American case history. pp. 467–480. *In* A. Lillehammer and S. J. Saltveit, eds., Regulated rivers. Universitetsforlaget, Oslo, Norway.

Stanford, J. A., and J. V. Ward. 1986. The Colorado River system. pp. 353–423. *In* B. R. Davies and K. F. Walker, eds., The ecology of river systems. Dr. W. Junk Publishers, Dordrecht, The Netherlands.

Stevens, L. 1983. The Colorado River in Grand Canyon: a comprehensive guide to its natural and human history, 5th ed. Red Lake Books, Flagstaff, Ariz.

Stevens, L. E., J. P. Shannon, and D. W. Blinn. 1997. Colorado River benthic ecology in Grand Canyon, Arizona: dam, tributary, and geomorphological influences. Regul. Rivers Res. Manag. 13: 129–149.

Stevens, L. E., J. E. Sublette, and J. P. Shannon. 1998. Chironomidae (Diptera) of the Colorado River, Grand Canyon, Arizona, USA, Part II: Factors influencing distribution. Great Basin Nat. 58(2): 147–155.

Stone, J. L., and N. L. Rathbun. 1967. Tailwater fisheries investigations, creel census and limnological study of the Colorado River below Glen Canyon Dam, July 1, 1966–June 30, 1967. Arizona Game and Fish Department, Phoenix, Ariz. 54 pp.

Stone, J. L., and N. L. Rathbun. 1969. Tailwater fisheries investigations, creel census and limnological study of the Colorado River below Glen Canyon Dam, July 1, 1968–June 30, 1969. Arizona Game and Fish Deptartment, Phoenix, Ariz. 60 pp.

Taba, S. S., J. R. Murphy, and H. H. Frost. 1965. Notes on the fishes in the Colorado River near Moab, Utah. Proc. Utah Acad. Sci. Arts Lett. 42(2): 280–283.

Tomko, D. S. 1975. The reptiles and amphibians of the Grand Canyon region. Plateau 47: 161–166.

Tyus, H. M. 1986. Life strategies in the evolution of the Colorado squawfish (*Ptychocheilus lucius*). Great Basin Nat. 46(4): 656–661.

Tyus, H. M., and C. A. Karp. 1989. Habitat use and streamflow needs of rare and endangered fishes, Yampa River, Colorado. U.S. Fish and Wildlife Service Biological Report 89(14). 27 pp.

U.S. Department of the Interior. 1997. Quality of water, Colorado River Basin Progress Report 18.

U.S. Environmental Protection Agency. 1971. The mineral quality problem in the Colorado River basin. Summary Report. U.S. EPA Regions VIII and IX.

U.S. Fish and Wildlife Service. 1988. The ecology of the lower Colorado River from Davis Dam to the Mexico–United States International Boundary: a community profile. U.S. Fish and Wildlife Service Biological Report 85(7.19). Sept. 1988.

U.S. Geological Survey. 1982. U.S. Geological Survey yearbook.

U.S. Geological Survey. 1995. Summary statistics and trend analysis of water-quality data at sites in the Gila River basin, New Mexico and Arizona. U.S. Geological Survey Water-Resources Investigations Report 95-4083.

U.S. Geological Survey. 1996. Physical and chemical characteristics of Lake Powell at the forebay and outflows of the Glen Canyon Dam, northeastern Arizona, 1990–1991. U.S. Geological Survey Water-Resources Investigations Report 96-4016.

Usher, H. D., and D. W. Blinn. 1990. Influence of various exposure periods on the biomass and chlorophyll *a* of *Cladophora glomerata* (Chlorophyta). J. Phycol. 26: 244–249.

Usher, H. D., et al. 1987. *Cladophora glomerata* and its diatom epiphytes in the Colorado River through Glen and Grand Canyons: distribution and desiccation tolerance. Final Report. Glen Canyon Environmental Studies. Bureau of Reclamation, Salt Lake City, Utah.

Vanicek, D. C., R. H. Kramer, and D. R. Franklin. 1970. Distribution of Green River fishes in Utah and Colorado following closure of Flaming Gorge Dam. Southwest. Nat. 14: 297–315.

Ward, J. V. 1975. Bottom fauna–substrate relationships in a northern Colorado trout stream: 1945 and 1974. Ecology 56(6): 1429–1434.

Ward, J. V. 1982. Altitudinal zonation of Plecoptera in a Rocky Mountain stream. Aquat. Insects 4(2): 105–110.

Ward, J. V. 1984. Stream regulation of the upper Colorado River: channel configuration and thermal heterogeneity. Int. Ver. Limnol. Verh. 22: 1862–1866.

Ward, J. V. 1986. Altitudinal zonation in a Rocky Mountain stream. Arch. Hydrobiol. Suppl. 74(2): 133–199.

Ward, J. V., and L. Berner. 1980. Abundance and altitudinal distribution of Ephemeroptera in a Rocky Mountain stream. pp. 169–177. *In* J. F. Flannagan, and K. E. Marshall, eds., Advances in Ephemeroptera biology. Plenum Press, New York.

Ward, J. V., and N. J. Voelz. 1988. Downstream effects of a reservoir on lotic zoobenthos. Int. Ver. Limnol. Verh. 23: 1174–1178.

Ward, J. V., and B. C. Kondratieff. 1992. An illustrated guide to the mountain stream insects of Colorado. Univ. Press of Colorado, Niwot, Colo.

Warren, P. L., and C. R. Schwalbe. 1985. Lizards along the Colorado River in Grand Canyon National Park: possible effects of fluctuating river flows. Glen Canyon Environmental Studies, Bureau of Reclamation, Upper Colorado Region, Salt Lake City, Utah. 19 pp.

Williams, W. D. 1987. Salinization of rivers and streams: an important environmental hazard. Ambio 16: 180–185.

Wydoski, R. C., K. Gilbert, K. Seethaler, C. W. McAda, and J. A. Wydoski. 1980. Annotated bibliography for aquatic resource management of the upper Colorado River ecosystem. U.S. Fish and Wildlife Service Resource Publication 135.

CHAPTER 2

Tributaries to the Colorado River

$$=========================$$

INTRODUCTION

Hardwater streams are found throughout Colorado. However, they are commonly found in the mountainous regions of the Colorado River basin, west of the Continental Divide. The basalt rock formations in this area are similar to those east of the divide in the softwater region: schist, gneisses, and granites. In the hardwater regions, however, these formations are overlain with soluble sedimentaries such as sandstone, limestone, and cretaceous shale (Pratt, 1938). Many underlying igneous rocks, varying from fine-grained to extremely coarse porphyritic granite and gneiss, are composed of large feldspar and mica pheocrysts in a quartz matrix (Knight, 1965). There are also site-specific increases in stream carbonate content resulting from the presence of hydrothermal marble (Knight and Argyle, 1962).

THE GUNNISON RIVER

The Gunnison River is formed by the junction of several headwaters particularly the North and South Gunnison. The South Gunnison River drainage (Figure 2.1) exemplifies hard headwater streams of the Colorado Rocky Mountain area. The South Gunnison originates west of the Continental Divide at the confluence of the East and Taylor Rivers in west-central Colorado at an altitude of 3283 m. It flows approximately 140 km to its confluence with the North Fork of the Gunnison River (2060 m). The Gunnison then continues to the Colorado River. The headwater tributaries to the Gunnison originate at altitudes of up to 4271 m. The Gunnison drains approximately

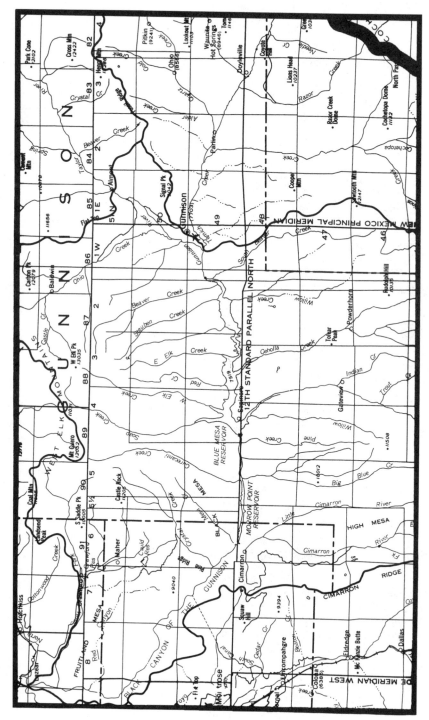

Figure 2.1. Gunnison River drainage, Colorado.

3885 km^2 (Knight and Gaufin, 1966) as it flows westward out of the Rocky Mountains. The South Gunnison is bounded by the West Elk Mountains on the north, the Elk Mountains on the northeast, the Sawatch Mountains on the east, Cochetopa Hills on the southeast, and the Uncompahgre Mountains on the south and the west (Figure 2.1) (Knight and Argyle, 1962).

Physical Characteristics

Flow in the Gunnison is typically quite rapid and turbulent. The discharge regime varies from minimum flows during autumn and winter, to spring maxima as a result of snowmelt in the headwaters. The lower river has been altered in some areas by channelization, impoundment, and diversion, and flow has become quite variable. Postregulation flow in these areas is considerably higher in winter and lower during spring, as runoff is stored in the reservoirs and discharged primarily from November to March. Over 90% of the average annual discharge is derived from precipitation in the headwaters. Reported current velocity values along the length of the river range widely, from 207 to 2597 m^3 s^{-1} (Stanford and Ward, 1983).

The river valley drops about 2.84 m km^{-1} above the Blue Mesa area. It continues into the Blue Mesa Reservoir and, farther downstream, the Morrow Point Reservoir. From there the river occupies a progressively narrowing gorge with exceedingly steep rocky slopes and high cliffs as it flows into the Black Canyon. The river drops about 5.68 m km^{-1} in this region, with swift rapids and frequent cascades.

The most important habitat in this area is clear rapids with stone substrata. Water flow is swift and bottoms are usually made up of well-rounded cobbles of varying size, with occasional larger rocks or boulders. During the 1940s and '50s some of this habitat has been eliminated by storage reservoirs, power developments, and irrigation practices. However, most of the clear, rocky riffle areas remain. The bottom of the Gunnison River proper is composed of large rubble and boulders. Sand and silt become more common in a downstream direction. Deep, slow-moving water and a sand and gravel bottom typify the junctions of the Gunnison and many of its tributaries (Argyle and Edmunds, 1962). Sediment loads that prior to impoundment were deposited in the lower river now collect in the reservoirs, leaving downstream discharges without any significant amount of suspended solids. The result of higher summer discharge and fewer suspended solids is a substrate in the Taylor River and the Black Canyon composed primarily of firmly embedded large rocks (Stanford and Ward, 1983). Turbidity in the South Gunnison is transitory and is primarily the result of erosion associated with snowmelt or a rainfall event.

The annual water temperature in the South Gunnison drainage ranges from a low of 0°C to a low-altitude summer high of about 26°C, although under natural conditions the maximum temperature in the Gunnison River proper is generally 20°C. The annual mean temperature of the river increases progressively downstream. The daily thermal gain (averaged over 12 months) between the headwaters and the Colorado River is about 6°C. In the Black Canyon, shading by the granite walls greatly influences the daily thermal regime; spring water temperatures are higher at night than during the day because of differential heating and cooling of the canyon walls (Figure 2.2). Hypolimnial release has resulted in an increase in winter temperatures

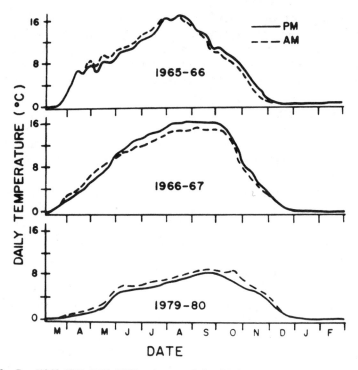

Figure 2.2.   Pre- (1965–1966, 1966–1967) and postregulation (1979–1980) temperature patterns measured 3 km downstream from Crystal Dam in the Gunnison River, Colorado. (From Stanford and Ward, 1983.)

and a 7 to 20°C decrease in summer temperatures below the dams (Stanford and Ward, 1983).

Chemical Characteristics

Hardness.   Pennak (1971b) has divided streams into three different hardness categories and gives the limits of these categories in mg L^{-1} as $CaCO_3$. Hardness can also be measured in ppm bound CO_2 (Table 2.1). The South Fork of the Gunnison River and its tributaries fall primarily into Pennak's "hard" and "very hard" stream classification categories, with total alkalinity values usually ranging from 130 to 220 mg L^{-1} as $CaCO_3$ (Table 2.2).

Alkalinity undergoes seasonal fluctuations as streams flood and recede. It also changes diurnally. A notable decrease occurs during the day as CO_2 and HCO_3 are absorbed. At night there is an increase in alkalinity as CO_2 is respired (Knight, 1965).

Dissolved Oxygen.   Dissolved oxygen (DO) values in the Gunnison River range from 7.4 to 9.5 ppm. The reported range of values for the Gunnison's hardwater tributaries is approximately 7.0 to 11.2 ppm. DO saturation of these systems typically varies between 90 and 115%. This situation is typical of mountain streams and can be attributed to the relative shallowness and significant surface exposure of these

TABLE 2.1. *Ranges of Values for Bound CO_2 Corresponding to Pennak's Three Categories*

	As $CaCO_3$ (mg L^{-1})	Bound CO_2 (ppm)
Medium	>60–120	>10–40
Hard	>120–180	>40–100
Very hard	>180	>100

Source: After Pennak (1971b).

streams, rapid current velocity, high degree of turbulence, and low water temperatures. These factors ensure that there is a constant and abundant supply of oxygen (Knight, 1965).

Hydrogen Ion Concentration. The waters of the Gunnison drainage are alkaline, due to the soluble sedimentary nature of the substrata over which the river flows. pH in the Gunnison River proper generally ranges from 7.8 to 8.6. The values are usually highest in the late summer when water levels are low and the dissolved solids are more concentrated. pH values are lowest in the spring when water levels are high and snowmelt and rainfall have had a dilution effect on dissolved solid concentrations (Knight, 1965).

Other Major Ion Concentrations. Concentrations of major ions in solution are highest in the downstream reaches of the river. Calcium is the dominant ion by percent composition above the Black Canyon, while sulfate loading from side flows draining gypsum formations characterize the lower river segment. Sulfate-containing salts were observed in high concentrations (e.g., >3000 mg L^{-1}) in the side flows (especially springbrooks and irrigation return flows) (Stanford and Ward, 1983).

TABLE 2.2. *Reported ppm $CaCO_3$ Values and/or Ranges of Values from Several South Gunnison Tributaries*

Tributary	$CaCO_3$ (ppm)
Cebolla Creek	159.0
Cement Creek	180.0
Cimarron Creek	185.8
Dry Gulch	123.0
East Elk Creek	222.0, 216.0
East River	116–192
Quartz Creek	152–196
Tomichi Creek	176, 144–188

Source: After Knight and Argyle (1962).

Conductivity.   Reported conductivity values for the Gunnison drainage range from 126 to 400 μS (Woodling, 1975).

Nutrients.   Nitrate (NO_3) values, reported as ppm N, are low in the Gunnison River drainage, usually varying between 0 and 1 ppm, with a mean of 0.19. Nitrate concentrations increase in a downstream direction over the river continuum. Nitrite (NO_2) in the Gunnison and its tributaries is usually negligible ppm as N. Ammonia (NH_3 as N ppm) is detectable throughout most of the drainage, although values are <1 ppm; the presence of NH_3 is probably due to irrigation return flows. Orthophosphate (as P ppm) is also occasionally detectable in the South Gunnison basin, however, these values are usually <1 ppm (Woodling, 1975).

This system appears to be nutrient limited at the headwaters, as are many cool mountain streams. Nutrient enrichment occurs, especially in areas with irrigation return flows. Although nonpoint sources of pollution appear to be primarily responsible for nutrification, domestic sewage treatment effluents also play a role in this process. Chemical and biological factors indicate that the effects of nutrient enrichment have been relatively minor and isolated throughout most of the Gunnison drainage (Richardson and Gaufin, 1971).

Total Dissolved Solids.   Reported total dissolved solids values for the Gunnison drainage range from 109 to 232 mg L^{-1}, with a maximum of 160 mg L^{-1} in the Gunnison River proper (Woodling, 1975).

Suspended Solids.   Reported suspended solids values for the South Gunnison drainage range from 3 to 11 mg L^{-1} (Woodling, 1975).

OTHER COLORADO HARD HEADWATER STREAMS

Physical Characteristics
There are a number of hard headwaters in Colorado, particularly those arising on the western slopes of the Rocky Mountains. Three major drainages occur partially in Colorado west of the Continental Divide: the Upper Colorado River Basin, of which the Gunnison River is a part, the Green River Basin, and the Dolores River Basin (Figure 2.3).

The streams in the Upper Colorado River Basin are predominantly hardwater streams. They usually occur in areas of high gradient, with rubble, gravel, and/or sand substrata (Holden and Stalnaker, 1975a). The Eagle River, the first large (above third order) tributary to the Colorado River, has hardness values falling in the medium-hard range (100 mg L^{-1} as $CaCO_3$) in May, during spring runoff, and values in the very hard range (270 mg L^{-1} as $CaCO_3$) in September (U.S. Geological Survey, 1982).

Roan Creek, a low-altitude tributary to Crystal Creek in the Upper Colorado basin, is a very hard water stream, with average bound CO_2 values of 220.8 ppm (Pennak, 1977) and hardness values averaging 500 mg L^{-1} as $CaCO_3$ (U.S. Geological Survey, 1982).

Crystal Creek flows northward out of the Elk Mountains from a 3950-m altitude. In the Montane regions bound CO_2 values average 39.8 ppm, in the medium-hard range (Pennak, 1977).

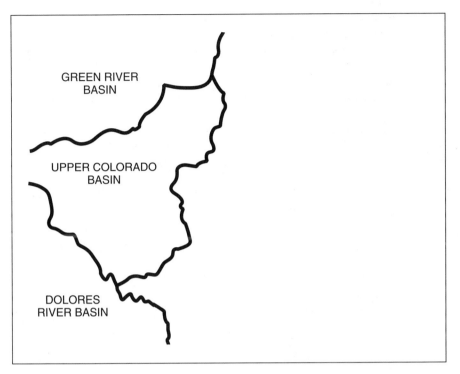

Figure 2.3.   Three major drainages west of the Continental Divide in Colorado.

Chemical Characteristics

Parachute Creek.   Parachute Creek is another very hard water stream of the upper Colorado drainage, with average bound CO_2 values of 178.0 ppm (Pennak, 1977) and average hardness values of 543 and 335 mg L^{-1} as $CaCO_3$ from two different sites (U.S. Geological Survey, 1982).

Green River System.   The Green River Basin is also largely a hardwater system. The White River, a major tributary to the Green River, is typically a hard to very hard water system with hardness values ranging from 110 to 350 mg L^{-1} as $CaCO_3$, spatially and temporally (U.S. Geological Survey, 1982). The White River basin is underlain by deposits of oil shale, halite, nahcolite, and gypsum. Annual precipitation ranges between 30 and 51 cm and temperature ranges from -40 to $40°C$. The headwaters are typically rubble and gravel riffles (Gray and Ward, 1979). The South Fork of the White River is also a hardwater stream with an average bound CO_2 value of 59.7 ppm (Pennak, 1977).

Green River Tributaries.   Two major tributaries to the White River (Green River drainage) are Piceance Creek and Yellow Creek. Piceance Creek is a very hard water stream with bound CO_2 values averaging 319.4 ppm (Pennak, 1977) and hardness as $CaCO_3$ ranging from 340 to 650 mg L^{-1} (U.S. Geological Survey, 1982). Yellow Creek is a very hard water stream with bound CO_2 values averaging 989.0 ppm

(Pennak, 1977) and hardness values ranging between 460 and 510 mg L^{-1} as $CaCO_3$ (U.S. Geological Survey, 1982).

The Yampa River system, another major tributary to the Green River, is primarily a medium-hard water system. Two of its headwater tributaries, Walton Creek and Trout Creek, fall into this category. Walton Creek has an average bound CO_2 value of 17.5 ppm (Pennak, 1977). Trout Creek has hardness values averaging 93 mg L^{-1} as $CaCO_3$ (Canton and Ward, 1981).

The Yampa River proper (tributary to Green River) has hardness values ranging from 50 mg L^{-1} in May to 260 mg L^{-1} as $CaCO_3$ (U.S. Geological Survey, 1982). It has low turbidity levels except for periods of short duration during snowmelt and rainfall runoff (Holden and Stalnaker, 1975b).

The Little Snake River, a lower-altitude tributary to the Green River, has spring runoff hardness values between 29 and 97 mg L^{-1} as $CaCO_3$; at other times, however, values range as high as 310 mg L^{-1} as $CaCO_3$ (U.S. Geological Survey, 1982). This is a relatively small, shallow, sandy river. However, it is extremely important to maintaining flows in the Yampa River (Holden and Stalnaker, 1975b).

Colorado River Tributaries.   The Dolores River basin, draining the southwestern portion of Colorado, appears to be a very hard water system. Its headwaters originate on the south slope of the San Miguel Mountains. The substrate is composed of large boulders, rubble, and coarse gravel. Peak discharge from snowmelt occurs from early May to late June (Fuller and Stewart, 1979). The San Miguel River, a major tributary to the Dolores River in this area, has hardness values ranging from 160 to 950 mg L^{-1} as $CaCO_3$ (U.S. Geological Survey, 1982).

ECOSYSTEM PROCESSES

Diatoms, unicellular and filamentous blue-green algae, unicellular green algae, small bits of leaves, conifer needles, and detritus are primarily responsible for primary production in Colorado hardwater streams. Because of the relative scarcity of large deciduous trees in the riparian vegetation, there is little accumulation of large leaf packs. Phytoplankton is sparse and of little importance in overall productivity; most suspended algae in these streams are periphytic cells sloughed from the substrata and swept into the current (Pennak, 1977). Allochthonous material is the most important source of organic nutrients (up to 90%) for bottom organisms in the Gunnison River (Pennak, 1977). Organic detritus appears to be a dominant factor in food webs and energy transfers from one trophic level to another in these streams (Pennak and Lavelle, 1979).

The total organic carbon pool in the Gunnison River increases from about 1.0 to 10.0 mg L^{-1}, on the average, from headwaters to mouth. Agglutination processes (i.e., demobilization of dissolved solids by conversion to particulate carbon forms) are responsible for progressively increasing particulate organic carbon (POC) values in a downstream direction. In lower river segments, side flows contribute significant amounts of allochthonous particulates. However, agglutination by autotrophic and microheterotrophic activity undoubtedly plays a major role in size fractions and POC concentrations, except during the spring freshet (Stanford and Ward, 1983).

In Piceance Creek, tributary to the White River (Green River drainage system), allochthonous material comprises 94% of the total sedimentary coarse particulate organic matter (CPOM) at 2103 m but only 27% at 1896 m. Standing crops of allochthonous material have been measured at 5.9 and 7.2 g dry weight m^{-2}, respectively. As with other Colorado hard headwaters, leaf litter is minor. Most allochthonous inputs are small bark and twig pieces, with less than 5% composed of leaf litter (Gray and Ward, 1979).

The overall composition of food ingested by Piceance Creek detritivore–herbivores attests to the importance of detritus in stream food webs: 73% detritus, 22% diatoms, and 5% filamentous algae. *Cladophora* sp. has a low energy content of 1.63 ± 0.68 kcal g^{-1} dry weight, while allochthonous organic matter has an energy content of 3.28 ± 0.46 kcal g^{-1} dry weight. CPOM has relatively little nutritive value until colonized by microorganisms (Gray and Ward, 1979).

In a northern Colorado springbrook (elevation 1597 m), detritus forms the major portion of the epilithon during much of the year (Ward and Dufford, 1979).

Although particulate organic matter (POC) in the South Gunnison River (Colorado drainage) increases naturally in a downstream direction, impoundment has altered this phenomenon somewhat. Accumulation of organic (and inorganic) matter in the reservoirs has resulted in low concentrations of particulate matter in dam outflows. As a result, POC concentrations are lower in reservoir tailwaters than in river segments above the reservoirs (Figure 2.4) (Stanford and Ward, 1983).

Benthic algae are the most important primary producers. *Nostoc, Oscillatoria, Rivularia, Stigonema, Ulothrix, Cladophora*, and diatoms comprise the bulk of the most commonly grazed autochthonous material. Little epilithic detritus or algae is present from May to August, as a result of high water scouring from snowmelt (Fuller and Stewart, 1979).

Net primary productivity, or *community metabolism*, in Colorado hard headwater streams is highly variable. Fluctuations between heterotrophic and autotrophic conditions occur over time at any given site. The occurrence of overall autotrophic conditions at one time and heterotrophic conditions at another time is possibly produced by changes in the amount of organic detritus and variations in the standing crop of lithophyton. There appear to be no seasonal or diurnal trends and no correlations between primary productivity values and light intensity or biomass of the bottom fauna (standing crop).

Despite the autotrophy–heterotrophy fluctuation, heterotrophic conditions are dominant, indicating that algal respiration and the action of bacteria and fungi on the organic detritus of the substrate more than balance algal photosynthesis. Microzoans and macrozoans of the substrate contribute further to this picture of heterotrophy, but to an unknown degree (Pennak and Lavelle, 1979). Also, sparse streamside vegetation and low local runoff appear to produce more stable autotrophic–heterotrophic ratios (Pennak and Lavelle, 1979).

Net primary productivity appears to be lower in these mountain headwater streams than in streams of lower altitudes. This is probably a result of the turbulent, vascillating environmental conditions that prevail in white-water habitats. Lower temperatures, poorly developed lithophyton, the possible presence of more organic detritus in

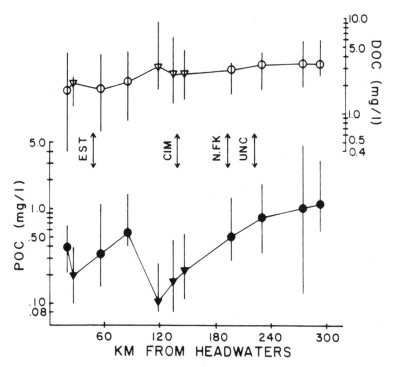

Figure 2.4.　Mean annual dissolved organic carbon (DOC) and particulate organic carbon (POC) concentrations (mg L^{-1} as C) measured at 11 sites on the Gunnison River. Inverted triangles indicate tailwater sites below mainstream dams; bars indicate range of values for 11 sampling periods during 1979 and 1980. Locations of major side flows are indicated by arrows. EST = East River, CIM = Cimarron River, N. FK = North Fork River, UNC = Uncompahgre River. (From Stanford and Ward, 1983.)

the rubble interstices, and the difficulty of measuring the rapid diffusion of oxygen and carbon dioxide through the air–water interface could all contribute to low productivity values (Pennak and Lavelle, 1979).

THE BIOTA OF COLORADO HARD HEADWATERS

See the species list (Table 2.3).

Riparian Vegetation

Riparian vegetation throughout most of the Gunnison drainage is composed of sagebrush (*Artemisia tridentata*), willow (*Salix* sp.), cottonwood (*Populus sargentii*), grasses, alder (*Alnus tenuifolia*), birch (*Betula occidentalis*), hawthorn (*Crataegus* sp.), wild rose, conifers, poison ivy (*Toxicodendron radicans*), serviceberry (*Amelanchier* sp.), dogwood (*Cornus* sp.), currant (*Ribes* sp.), and raspberry (*Rubus* sp.) (Knight, 1965; Knight and Argyle, 1962). In the Black Canyon Gorge area the banks are steep cliffs with little or no vegetation.

(text continues on page 97)

TABLE 2.3. *Species List: Colorado River Hard Headwaters, Gunnison Drainage*

Taxon[a]	Substrate[b]					Feeding Habits[c]
	V	S	M	G	C	
SUPERKINGDOM PROKARYOTAE						
KINGDOM MONERA						
Division Cyanophycota						
Class Cyanophyceae						
Order Nostocales						
Family Chamaesiphonaceae						
Chamaesiphon incrustans	×					
Family Nostocaceae						
Anabaena sp.					×	
Nostoc parmeloides					×	
N. verrucosum					×	
Nostoc. sp.					×	
Family Oscillatoriaceae						
Lyngbya sp.						
Oscillatoria princeps					×	
Oscillatoria sp.					×	
Phormidium sp.					×	
Family Rivulariaceae						
Gloeotrichia sp.						
Rivularia sp.					×	
Family Stigonemataceae						
Stigonema sp.					×	
SUPERKINGDOM EUKARYOTAE						
KINGDOM PLANTAE						
Subkingdom Thallobionta						
Division Chlorophycota						
Class Chlorophyceae						
Order Chlorococcales						
Family Characiaceae						
Characium sp.	×				×	
Family Hydrodictyaceae						
Hydrodictyon sp.		×	×			
Pediastrum sp.		×	×			
Family Micractiniaceae						
Golenkinia sp.		×	×			
Micractinium sp.		×	×			
Family Oocystaceae						
Ankistrodesmus sp.			×			
Family Scenedesmaceae						
Scenedesmus sp.					×	
Order Volvocales						
Family Chlamydomonadaceae						
Chlamydomonas sp.	×	×	×			
Order Ulotrichales						
Family Chaetophoraceae						
Chaetophora sp.					×	
Microthamnion sp.					×	
Stigeoclonium sp.					×	
Family Microsporaceae						
Microspora sp.					×	
Family Ulotrichasceae						
Ulothrix zonata					×	

(continues)

TABLE 2.3. *Continued*

Taxon[a]	Substrate[b]					Feeding Habits[c]
	V	S	M	G	C	
Ulothrix spp.					×	
Order Zygnematales						
Family Desmidiaceae						
Closterium cucumis			×		×	
Closterium sp.	×		×		×	
Cosmarium undulatum	×		×		×	
Cosmarium sp.	×		×		×	
Euastrum sp.	×		×		×	
Micrasterias sp.	×		×		×	
Family Zygnemataceae						
Mougeotia sp.					×	
Spirogyra sp.					×	
Zygnema sp.					×	
Order Cladophorales						
Family Cladophoraceae						
Cladophora glomerata	×				×	
Cladophora sp.	×				×	
Rhizoclonium sp.	×				×	
Order Ulvales						
Family Ulvaceae						
Enteromorpha sp.					×	
Order Tetrasporales						
Family Tetrasporaceae						
Tetraspora sp.						
Order Oedogoniales						
Family Oedogoniaceae						
Oedogonium sp.					×	
Order Schizogoniales						
Family Schizogoniaceae						
Prasiola sp.					×	
Class Bacillariophyceae						
Order Eupodiscales						
Family Coscindodiscaceae						
Cyclotella sp.						
Melosira varians					×	
Melosira sp.					×	
Order Fragilariales						
Family Fragilariaceae						
Asterionella formosa						
Asterionella sp.						
Diatoma hiemale					×	
D. hiemale var. *hiemale*					×	
D. hiemale var. *mesodon*					×	
D. tenue var. *elongatum*					×	
D. vulgare					×	
Diatoma sp.					×	
Fragilaria brevistriate					×	
F. capucine					×	
F. construens					×	
F. crotonensis					×	
F. leptostauron					×	
F. pinnata var. *lancettula*					×	
F. vaucheriae					×	

TABLE 2.3. *Continued*

Taxon[a]	Substrate[b]					Feeding Habits[c]
	V	S	M	G	C	
Fragilaria spp.					×	
Hannaea arcus						
Meridion circulare					×	
Meridion sp.					×	
Opephora martyi						
Synedra parasitica var. *subconstricta*	×					
S. rumpens	×					
S. ulna	×					
Synedra spp.	×		×		×	
Tabellaria sp.						
Order Eunotiales						
Family Eunotiaceae						
Actinella sp.						
Eunotia curvata						
Order Achnanthales						
Family Achnanthaceae						
Achnanthes affinis					×	
A. clevei					×	
A. lanceolata					×	
A. linearis					×	
A. minutissima					×	
A. saxonica					×	
Achnanthes sp.					×	
Cocconeis pediculus					×	
C. placentula					×	
Cocconeis sp.					×	
Rhoicosphenia curvata						
Order Naviculales						
Family Cymbellaceae						
Amphora ovalis						
A. perpusilla						
Cymbella cistula					×	
C. minuta					×	
C. prostrata					×	
C. pusilla					×	
C. sinuata					×	
C. tumida					×	
Cymbella spp.					×	
Family Gomponemaceae						
Gomphonema acuminatum					×	
G. angustatum					×	
G. brebissonii					×	
G. clevei					×	
G. ehrengregii					×	
G. herculeana var. *robusta*					×	
G. olivaceiodes					×	
G. olivaceum					×	
G. parvulum					×	
G. subclavatum					×	
G. truncatum					×	
Gomphonema spp.					×	
Family Naviculaceae						
Anomoeoneis vitrea						

(continues)

TABLE 2.3. *Continued*

Taxon[a]	Substrate[b]					Feeding Habits[c]
	V	S	M	G	C	
Caloneis lewigii						
C. ventricosa var. *alpina*						
Caloneis sp.						
Diatomella sp.						
Diploneis elliptica						
Frustulia rhomboides						
Frustulia sp.						
Gyrosigma acuminatum						
Gyrosigma sp.						
Navicula bacillum					×	
N. cryptocephala					×	
N. cuspidata					×	
N. elginensis					×	
N. exigua					×	
N. exigua var. *capitata*					×	
N. minima					×	
N. parvulum					×	
N. pelliculosa					×	
N. pupula					×	
N. pupula var. *capitata*					×	
N. radiosa					×	
N. rhynchocephala					×	
N. salinarum					×	
N. tripunctata					×	
N. viridula					×	
Navicula spp.					×	
Neidium binode						
Pinnularia intermedia						
P. rupestris						
Pinnularia sp.						
Stauroneis smithii						
Stauroneis sp.						
Order Epithemiales						
Family Epithemiaceae						
Denticula elegans						
Epithemia gibba					×	
E. sorex					×	
E. turgida					×	
Order Bacillariales						
Family Nitzschiaceae						
Nitzschia acicularis					×	
N. amphibia					×	
N. dissipata					×	
N. fonticola					×	
N. hungarica					×	
N. linearis					×	
N. palea					×	
N. sigmoidea					×	
Nitzschia sp.					×	
Order Surirellales						
Family Surirellaceae						
Surirella angustata						
S. ovalis						

TABLE 2.3. *Continued*

Taxon[a]	Substrate[b]					Feeding Habits[c]
	V	S	M	G	C	
S. ovata						
Division Chromophycota						
Class Xanthophyceae						
Order Heterosiphonales						
Family Vaucheriaceae						
Vaucheria sp.					×	
Class Chrysophyceae						
Order Chrysocapsales						
Family Hydruraceae						
Hydrurus foetidus					×	
Hydrurus sp.					×	
Division Euglenophyta						
Class Euglenophyceae						
Order Euglenales						
Family Euglenaceae						
Euglena sp.	×		×		×	
Phacus sp.	×		×		×	
Trachelomonas sp.						
Division Pyrrophyta						
Class Dinophyceae						
Order Peridinales						
Family Ceratiaceae						
Ceratium hirudinella						
Family Peridiniaceae						
Peridinium sp.						
Division Rhodophyta						
Class Rhodophyceae						
Order Nemalionales						
Family Batrachospermaceae						
Batrachospermum monoliforme					×	
Division Charophyta						
Class Charophyceae						
Order Charales						
Family Characeae						
Chara sp.						
Division Bryophyta						
Class Musci						
Amblystegium sp.					×	
Scleropodium obtusifolium					×	
Scleropodium sp.					×	
Division Magnoliophyta						
Class Monocotyledonae						
Order Najadales						
Family Potamogetonaceae						
Potamogeton crispus		×	×			
P. filiformis		×	×			
Potamogeton sp.		×	×			
Family Zannichelliaceae						
Zannichellia sp.						
Order Hydrocharitales						
Family Hydrocharitaceae						
Elodea sp.						
Order Cyperales						
Family Cyperaceae						

(continues)

TABLE 2.3. *Continued*

Taxon[a]	Substrate[b] V	S	M	G	C	Feeding Habits[c]
Carex sp.						
Eleocharis sp.						
Scirpus sp.						
Order Typhales						
Family Typhaceae						
Typha sp.						
Order Arales						
Family Lemnaceae						
Lemna minor						
Class Dicotyledonae						
Order Ranunculales						
Family Ranunculaceae						
Ranunculus aquatilis		×	×			
Ranunculus sp.		×	×			
Order Haloragales						
Family Haloragaceae						
Myriophyllum sp.						
Order Caprales						
Family Brassicaceae						
Rorippa nasturtium-aquaticum						
Rorippa sp.						
KINGDOM ANIMALIA						
Subkingdom Protozoa						
Class Mastigophora						
Order Cryptomonadida						
Family Cryptomonadidae						
Chilomonas paramecium						O
Cryptomonas sp.						O
Class Ciliata						
Order Holotrichida						
Family Amphileptidae						
Loxophyllum setigera						O
Family Frontoniidae						
Frontonia leucas						O
Family Holophryidae						
Chaenia teres						O
Family Paramecidae						
Paramecium sp.						O
Order Peritrichida						
Family Vorticellidae						
Vorticella sp.					×	O
Order Protomastigida						
Family Protomastigineae						
Bodo caudatus						O
Order Spirotrichida						
Family Bursariidae						
Bursaria truncatella						O
Family Metopidae						
Metopus sigmoides						O
Family Oxytrichidae						
Holosticha vernalis						O
Stylonychia notophora						O

TABLE 2.3. *Continued*

Taxon[a]	Substrate[b]					Feeding Habits[c]
	V	S	M	G	C	
Family Spirostomidae						
Blepharisma lateritia						O
Class Actinopoda						
Order Actinophrydia						
Actinosphaerium sp.						O
Class Lobosa						
Order Testacealobosa						
Family Centropyxidae						
Centropyxis spp.						O
Family Difflugiidae						
Difflugia sp.						O
Order Amoebaea						
Family Chaosidae						
Amoeba limax			×			O
A. proteus			×			O
Amoeba sp.			×			O
Phylum Rotifera						
Class Monogononta						
Order Ploima						
Family Brachionidae						
Brachionus calyciflorus						O
Brachionus sp.						O
Keratella cochlearis						O
K. quadrata						O
Notholca sp.						O
Family Lecanidae						
Monostyla lunaris						O
Family Synchaetidae						
Polyarthra trigla						C
Polyartha sp.						C
Synchaeta sp.						C
Order Flosculariaceae						
Family Testudinellidae						
Filinia longiseta						O
Phylum Platyhelminthes						
Class Turbellaria						
Order Tricladida						
Family Planariidae						
Dugesia dorotocephala	×	×	×	×	×	D
Dugesia sp.	×	×	×	×	×	D
Planaria sp.	×	×	×	×	×	D
Phylum Nematomorpha						
Order Gordiida						
Family Chordodidae						
Paragordius varius						C
Phylum Annelida						
Class Oligochaeta						
Order Lumbriculida						
Family Lumbricidae						
Eiseniella tetraedra		×	×	×	×	HD
E. tetraedra hercynia		×	×	×	×	HD
E. tetraedra typica		×	×	×	×	HD
Helodrilus sp.				×		HD

(continues)

TABLE 2.3. *Continued*

Taxon[a]	Substrate[b]					Feeding Habits[c]
	V	S	M	G	C	
Order Haplotaxida						
Family Enchytraeidae						
Family Naididae						
Paranais sp.		×	×			HD
Pristina sp.		×	×			HD
Family Tubificidae						
Limnodrilus hoffmeisteri						HD
Tubifex tubifex						HD
Class Hirudinaceae						
Order Rhynchobdellida						
Family Glossiphoniidae						
Batracobdella picta						C
Glossiphonia complanata					×	C
Hellobdella stagnalis					×	C
Placobdella ornata		×	×	×		C
Theromyzon rude					×	C
Order Pharyngobdellida						
Family Erpobdellidae						
Dina dubia	×					C
D. fervida	×					C
D. microstoma						C
D. parva						C
Erpobdella punctata			×	×	×	C
Nephelopsis obscura						C
Order Gnathobdellida						
Family Hirudidae						
Haemopsis marmorata		×				C
Phylum Arthropoda						
Subphylum Mandibulata						
Class Crustacea						
Subclass Malacostraca						
Order Amphipoda						
Family Gammaridae						
Gammarus lacustris				×		O
Gammarus sp.				×		O
Stygobromus coloradensissim						O
S. holsingeri						O
S. pennaki						O
Family Talitridae						
Hyallela azteca						O
Subclass Eucopepoda						
Order Copepoda						
Family Cyclopidae						
Cyclops bicuspidatus						OD
C. bicuspidatus thomasi						OD
Family Diaptomidae						
Diaptomus nudus						OD
Diaptomus sp.						OD
Subclass Branchiopoda						
Order Cladocera						
Family Bosminidae						
Bosmina coregoni						OD
B. longirostris						OD
Bosmina sp.						OD

TABLE 2.3. *Continued*

Taxon[a]	V	S	M	G	C	Feeding Habits[c]
Family Chydoridae						
Alona sp.						OD
Chydorus sp.	×					OD
Family Daphnidae						
Ceriodaphnia quadrangula						OD
Daphnia longispina						OD
Daphnia sp.						OD
Simocephalus sp.	×					OD
Class Insecta						
Order Coleoptera						
Family Chrysomelidae						
Donacia sp.						H
Family Dryopidae						
Helichus sp.						O
Family Dytiscidae						
Deronectes sp.					×	C
Laccophilus terminalis						C
Rhantus gutticolis						C
Family Elmidae						
Heterlimnus sp.	×				×	O
Narpus sp.						O
Optioservus seriatus					×	O
Optioservus sp.	×				×	O
Zaitzevia parvula					×	O
Zaitzevia sp.					×	O
Family Haliplidae						
Peltodytes sp.						C
Family Heteroceridae						
Heterocerus sp.						O
Family Hydrophilidae						
Enochrus diffuses						HO
Laccobius carri						H
Tropisternus lateralis						O
Order Collembola						
Family Isotomidae						
Isotomurus palustris						O
Order Diptera						
Family Blephariceridae						
Agathon sp.						O
Bibiocephala sp.						O
Family Chironomidae						
Ablabesmyia sp.						CO
Calopsectra sp.						O
Cardiocladius sp.					×	C
Chironomus plumosus gr.						OH
C. staegeri gr.						OH
C. thummi gr.						OH
Chironomus spp.						OH
Conchapelopia sp.						C
Cricotopus trifasciatus gr. sp.						O
Cricotopus spp.						O
Cricotopus/Orthocladius complex						O
Cryptochironomus sp.						O
Diamesa sp.					×	O

(continues)

TABLE 2.3.   *Continued*

Taxon[a]	Substrate[b]					Feeding Habits[c]
	V	S	M	G	C	
Dicrotendipes modestus						OH
Eukiefferiella spp.					×	OC
Glyptotendipes sp.						OH
Limnophora sp.						O
Odontomesa fulva						O
Orthocladius sp.					×	HD
Palpomyia sp.						C
Pentaneura sp.						CO
Polypedilum sp.						OC
Potomyria sp.						
Procladius sp.					×	OC
Prodiamesa olivacea						O
Prodiamesa sp.		×				O
Psectrocladius sp.						OH
Pseudochironomus sp.						O
Pseudodiamesa pertinax					×	O
Pseudodiamesa spp.					×	O
Rheotanytarsus sp.						O
Stictochironomus sp.		×				O
Syndiamesa sp.						O
Tanypus sp.						CO
Family Deuterophlebiidae						
Deuterophlebia sp.					×	O
Family Empididae						
Hemerodromia sp.						CO
Family Muscidae						
Limnophora sp.					×	C
Family Psychodidae						
Pericoma sp.	×				×	O
Family Rhagionidae						
Atherix variegata					×	C
Atherix spp.					×	C
Family Simuliidae						
Prosimulium sp.						O
Simulium arcticum					×	O
S. hunteri					×	O
S. latipes					×	O
S. vittatum					×	O
Simulium spp.					×	O
Family Stratiomyidae						
Euparyphus sp.						O
Family Tanyderidae						
Protoplasma sp.					×	O
Family Tipulidae						
Antocha sp.						O
Hexatoma sp.					×	C
Limonia sp.	×					H
Tipula commiscibilis					×	O
Tipula spp.					×	O
Order Ephemeroptera						
Family Baetidae						
Baetis bicaudatus					×	DO
B. tricaudatus					×	DO

TABLE 2.3. *Continued*

Taxon[a]	Substrate[b] V	S	M	G	C	Feeding Habits[c]
Baetis spp.					×	DO
Callibaetis nigritus					×	O
Callibaetis sp.					×	O
Centroptilum sp.					×	DO
Paracloeodes sp.					×	O
Pseudocloeon sp.					×	O
Family Caenidae						
Caenis sp.						O
Family Ephemerellidae						
Ephemerella coloradensis					×	O
E. doddsi					×	O
E. fuscata					×	O
E. grandis					×	O
E. grandis grandis					×	O
E. hecuba hecuba					×	O
E. inermis					×	O
E. infrequens					×	O
E. margarita					×	O
E. needhami					×	O
E. tibialis	×				×	O
Ephemerella spp.					×	O
Family Heptageniidae						
Cinygma sp.					×	O
Cinygmula spp.					×	O
Epeorus albertae					×	O
E. longimanus					×	O
Epeorus spp.					×	O
Heptagenia elegantula					×	O
H. solitaria					×	O
Heptagenia spp.		×			×	O
Rhithrogena doddsi						OD
R. hageni					×	OD
R. robusta					×	OD
Rhithrogena spp.					×	OD
Stenonema sp.						O
Family Leptophlebiidae						
Paraleptophlebia pallipes			×		×	DO
P. vaciva			×		×	DO
Paraleptophlebia spp.					×	DO
Family Siphlonuridae						
Ameletus velox					×	OD
Ameletus spp.					×	OD
Siphlonurus occidentalis					×	OC
Siphlonurus sp.					×	OC
Family Tricorythidae						
Tricorythodes fallax					×	O
T. minutus					×	O
Tricorythodes sp.					×	O
Order Hemiptera						
Family Corixidae						
Cenocorixa utahensis						O
Trichocorixa calva						C

(continues)

TABLE 2.3. *Continued*

Taxon[a]	Substrate[b]					Feeding Habits[c]
	V	S	M	G	C	
Family Gerridae						
Gerris comatus						C
G. remigis						C
Order Odonata						
Family Coenagrionidae						
Argia lugens						C
Ishnura spp.						C
Family Gomphidae						
Ophiogomphus severus						C
Ophiogomphus sp.						C
Order Plecoptera						
Family Capniidae						
Capnia coloradensis					×	D
C. confusa					×	D
C. gracilaria					×	D
C. limita					×	D
C. logana					×	D
C. poda					×	D
Eucapnopsis brevicauda					×	D
Family Chloroperlidae						
Alloperla borealis					×	OC
A. coloradenis					×	OC
A. pallidula					×	OC
A. pintada					×	OC
A. signata			×		×	OC
Alloperla spp.					×	OC
Kathroperla sp.					×	O
Paraperla frontalis					×	O
Paraperla sp.					×	O
Suwallia sp.					×	O
Sweltsa sp.					×	O
Triznaka signata					×	O
Triznaka sp.					×	O
Family Nemouridae						
Amphinemura sp.					×	DO
Brachyptera pacifica					×	D
B. pallida					×	D
Isocapnia crinita					×	D
Leuctra sara					×	D
Malenka flexura					×	D
Nemoura besametsa					×	D
N. cinctipes					×	D
N. coloradensis					×	D
N. oregonensis					×	D
Nemoura spp.					×	D
Paraleuctra sp.					×	D
Podmosta sp.					×	D
Prostoia besametsa					×	D
Taeniopteryx sp.					×	D
Zapada cinctipes					×	D
Z. haysi					×	D
Family Perlidae						
Acroneuria depressa					×	C
A. pacifica					×	C

TABLE 2.3. *Continued*

Taxon[a]	Substrate[b]					Feeding Habits[c]
	V	S	M	G	C	
Acroneuria spp.					×	C
Atoperla sp.					×	C
Claassenia sabulosa					×	C
Claasenia sp.					×	C
Hesperoperla pacifica					×	C
Neoperla sp.					×	C
Family Perlodidae						
Arcynopteryx paralella					×	C
A. signata					×	C
Arcynopteryx sp.					×	C
Cultus aestivalis					×	C
Diura knowltoni					×	C
Isogenoides zionensis					×	C
Isogenus aestivalis					×	C
I. elongatus					×	C
I. expansus					×	C
I. modestus					×	C
Isogenus sp.					×	C
Isoperla bilineata					×	C
I. ebria					×	C
I. fulva					×	C
I. mormona					×	C
I. patricia					×	C
I. pinta					×	C
Isoperla sp.					×	C
Kogotus modestus					×	C
Megarcys signata					×	C
Skwala parallela					×	C
Family Pteronarcidae						
Pteronarcella badia			×			OC
Pteronarcella sp.			×			OC
Pteronarcys californica	×				×	OC
Pteronarcys sp.	×				×	OC
Order Trichoptera						
Family Brachycentridae						
Brachycentrus americanus					×	O
B. similis					×	O
Brachycentrus spp.					×	O
Micrasema sp.					×	OH
Family Glossosomatidae						
Agapetus boulderensis					×	O
Agapetus sp.					×	O
Glossosoma parvulum					×	O
G. ventrale					×	O
Glossosoma spp.					×	O
Family Hydropsychidae						
Arctopsyche grandis					×	O
A. inermis					×	O
Arctopsyche spp.					×	O
Cheumatopsyche sp.					×	O
Hydropsyche cockerelli					×	O
H. morosa					×	O
H. oslari					×	O

(continues)

TABLE 2.3. *Continued*

Taxon[a]	Substrate[b]					Feeding Habits[c]
	V	S	M	G	C	
Hydropsyche spp.					×	O
Micronemum sp.					×	O
Parapsyche sp.					×	O
Family Hydroptilidae						
Agraylea sp.		×	×	×	×	HO
Hydroptila spp.		×	×	×	×	HO
Stactobiella sp.		×	×	×	×	O
Family Lepidostomidae						
Lepidostoma moneka					×	D
Lepidostoma spp.					×	D
Family Leptoceridae						
Leptocella sp.						O
Family Limnephilidae						
Ecclisomyia conspersa						
Hesperophylax consimilis					×	O
H. incisus	×		×	×	×	HD
Hesperophylax sp.					×	HD
Limnephilus frijole			×	×	×	HD
Neothremma alicia					×	O
Oligophlebodes minutum					×	D
Oligophlebodes sp.					×	O
Family Psychomyiidae						
Psychomyia pulchella					×	O
Family Rhyacophiloidea						
Rhyacophila acropedes					×	OC
R. alberta					×	OC
R. angelita					×	OC
R. coloradensis					×	OC
R. hyalinata					×	OC
R. valuma					×	OC
R. verrula					×	OC
Rhyacophila spp.					×	OC
Family Sericostamatidae						
Sericostoma sp.						DO
Phylum Mollusca						
Class Gastropoda						
Order Basommatophora						
Family Ancylidae						
Ancylus sp.					×	HD
Family Lymnaeidae						
Lymnaea spp.		×	×	×	×	O
Family Physidae						
Physa anatina	×	×	×	×	×	O
Physa sp.	×	×	×	×	×	O
Family Planorbidae						
Gyraulus sp.			×	×		HD
Class Pelecypoda						
Order Heterodonta						
Family Sphaeriidae						
Pisidium casertanum sp.			×			D
P. compressum			×			D
Pisidium sp.		×	×	×	×	D
Phylum Chordata						
Subphylum Vertebrata						

TABLE 2.3. *Continued*

Taxon[a]	Substrate[b]					Feeding Habits[c]
	V	S	M	G	C	
Class Osteichthyes						
Order Atheriniformes						
Family Cyprinodontidae						
Fundulus kansae						C
F. sciadicus						C
F. zebrinus						C
Order Salmoniformes						
Family Esocidae						
Esox lucius					×	C
Family Salmonidae						
Prosopium williamsoni	×					C
Salmo clarki				×	×	C
S. clarki pleuriticus						C
S. clarki stomias						C
S. gairdneri				×	×	C
S. trutta				×	×	C
Salvelinus fontinalis						C
Thymallus articus						C
Trutta fario						C
T. fario levenensis						C
T. pleuriticus						C
T. shasta						C
Order Cypriniformes						
Family Catostomidae						
Carpiodes carpio carpio						C
C. cyprinus						C
Catostomus catostomus						C
C. catostomus griseus						C
C. commersoni					×	C
C. discobolus					×	C
*C. discobolus × C. commersoni					×	C
C. latipinnis					×	C
*C. latipinnis × C. commersoni					×	C
*C. latipinnis × Xyrauchen texanus					×	C
Moxostoma macrolepidotum						C
Pantosteus delphinus						C
P. platyrhynchus						C
Xyrauchen cypho						C
X. texanus				×	×	C
Family Cyprinidae						
Agosia yarrowi						C
Apocope yarrowi						C
Campostoma anomalum						O
C. anomalum pullum						O
Carassius auratus						O
Chrosomus eos						H
Couesius plumbea						H
Cyprinus carpio	×		×		×	O
Gila atraria						O
G. cypha						O
G. cypha complex				×	×	O
G. elegans						O
G. robusta				×	×	O
Hybognathus hankinsoni				×	×	D

(continues)

TABLE 2.3. *Continued*

Taxon[a]	Substrate[b]					Feeding Habits[c]
	V	S	M	G	C	
H. nuchalis						D
H. placitus						D
Hybopsis biguttata						C
H. storeiana						C
Notropis blennius						C
N. cornutus						CO
N. dorsalis						CO
N. dorsalis piptolepis						CO
N. heterolepis						CO
N. lutrensis				×		CO
N. lutrensis lutrensis						CO
N. stramineus				×		CO
N. stramineus missuriensis						CO
Phenacobius mirabilis						C
Pimephales promelas	×				×	DH
Ptychocheilus lucius					×	C
Rhinichthys cataractae						C
R. cataractae dulcis						C
R. cataractae pullum						C
R. osculus				×	×	C
Richardsonius balteatus					×	O
Semotilus atromaculatus						C
Order Perciformes						
Family Centrarchidae						
Lepomis cyanellus	×				×	C
L. gibbosus						C
L. humilis						C
L. macrochirus					×	C
Micropterus salmoides					×	C
Pomoxis annularis						C
P. nigromaculatus						C
Family Cottidae						
Cottus bairdi					×	C
C. bairdi punctulatus						C
Family Percidae						
Etheostoma exile						C
E. nigrum						C
E. nigrum nigrum						C
E. spectabile pulchellum						C
Perca flavescens						C
Stizostedion vitreum					×	C
Order Siluriformes						
Family Ictaluridae						
Ictalurus melas			×		×	C
I. nebulosus						C
I. punctatus			×		×	C
Noturus flavus						C
Order Gasterosteiformes						
Family Gasterosteidae						
Culaea incostans						C

[a]An asterisk indicates an introduced species.
[b]V, Vegetation and debris; S, sand; M, mud; G, gravel; C, cobble/rubble.
[c]Feeding habits: C = carnivore; CO = carnivore–omnivore; D = detritivore; DH = detritivore–herbivore;
DO = detritivore–omnivore; H = herbivore; HD = herbivore–detritivore; HO = herbivore–omnivore;
O = omnivore; OC = omnivore–carnivore; OD = omnivore–detritivore; OH = omnivore–herbivore.

Algae

A blue-green alga (Cyanophyta), *Nostoc parmeloides*, grows on stones in running water, in the neighborhood of 2850 m altitude. The seasonal decrease in water temperature in early August seems to trigger growth of the macroscopic *Nostoc parmeloides* colonies; they reach their peak abundance in late February and early March, at which time they frequently completely cover the substrata in swiftly flowing riffle areas. Midge larvae are often found associated with these *Nostoc* colonies, and developing larvae consume the bulk of algal material. The remainder of the colony breaks up soon after emergence of the insect, although small portions persist throughout the summer. Analyses of trout intestinal contents have revealed that these *Nostoc* colonies are often a major source of food, frequently equaling or surpassing volumes of insects found in trout stomachs. *Nostoc verrucosum* is also abundant on rocks in the South Gunnison (Richardson and Gaufin, 1971), and *Nostoc* sp. has been known to occur in marginal pools in quiet water, between 2256 and 2408 m altitude (Pratt, 1938).

Oscillatoria sp. is frequently found in the Gunnison River as characteristic slimy masses on rock surfaces in late summer. Although these colonies wash away rather easily, numerous chironomid larvae are often found associated with them (Pratt, 1938).

The blue-green algae of a northern Colorado springbrook (elevation 1597 m) exhibit significant seasonal fluctuations, with peaks during June and July. *Chamaesiphon incrustans* occurs epiphytically on *Cladophora glomerata*. *Anabaena* sp., *Phormidium* sp., and *Oscillatoria princeps* are also present (Ward and Dufford, 1979).

Euglena sp. and *Phacus* sp. are the only Euglenophyta in a northern Colorado springbrook (elevation 1597 m) (Ward and Dufford, 1979).

The most common green algae (Chlorophyceae) in the South Gunnison River drainage are *Cladophora* sp., *Stigeoclonium* sp., and *Ulothrix* spp. (Pratt, 1938).

Cladophora sp. is the most commonly observed alga in the lower South Gunnison River. It is perennial and usually reaches maximum growth during August and September, when it appears as long plumes in swiftly flowing rapids at moderate depths. In late September these filaments are washed away, and a sparse mat remains until growth resumes the following summer. *Cladophora* is found in the Gunnison only below an altitude of 2347 m. Its distribution coincides remarkably with that of the stonefly *Pteronarcys californica*. *Ulothrix* sp. is found occasionally in deposits between rocks, and in marginal pools between 2256 and 2408 m elevation (Pratt, 1938).

The desmids *Euastrum* sp. and *Cosmarium undulatum* are rare in the Gunnison and generally found mixed with detritus or attached to other aquatic vegetation. *Closterium* sp. occurs sparingly on rocks in shallow rapids between 2256 and 2408 m altitude (Pratt, 1938).

Spirogyra sp. is confined to recessional pools and rarely spreads into the main stream (Pratt, 1938). *Prasiola* sp. is also present and is found to be fairly common on rocks in shallow water. *Mougeotia* sp., *Oedogonium* sp., *Spirogyra* sp., *Stigonema* sp., and *Microspora crassior* have all been observed in gut contents of some western stonefly nymphs (Richardson and Gaufin, 1971).

The algal communities in Piceance Creek vary seasonally and altitudinally (Figures 2.5 and 2.6). *Cladophora glomerata* comprises 13% of the total periphyton and 6% of the total CPOM at 2103 m. At 1896 m, *Cladophora* is abundant and occurs along with *Enteromorpha* sp., *Vaucheria* sp., and *Rivularia* sp. From May through October, filamentous algae account for 63% of the total periphyton and 60% of the total CPOM (Gray and Ward, 1979).

Algal populations in a northern Colorado springbrook (elevation 1597 m) undergo

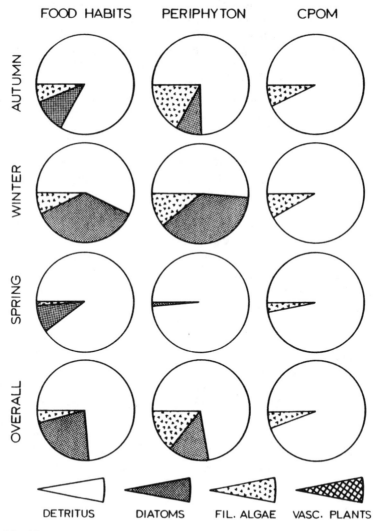

Figure 2.5.   Mean percentage composition of major dietary components of herbivore–detritivores compared with the composition of these components in periphyton and coarse particulate organic matter (CPOM) at an altitude of 2103 m in Piceance Creek, Colorado. (From Gray and Ward, 1979.)

significant seasonal fluctuations (Figure 2.7). Chlorophytes dominate the epilithon throughout most of the year. *Cladophora glomerata* is the most abundant and widespread green alga and has been collected year-round, except during February and March, with maximum growth occurring in October and November. *Ulothrix zonata* is also relatively common and exhibits fall and winter maxima. *Characium* sp., *Chaetophora* sp., *Microspora* sp., *Oedogonium* sp., *Spirogyra* sp., *Stigeoclonium* sp., and *Zygnema* sp. are also present (Ward and Dufford, 1979).

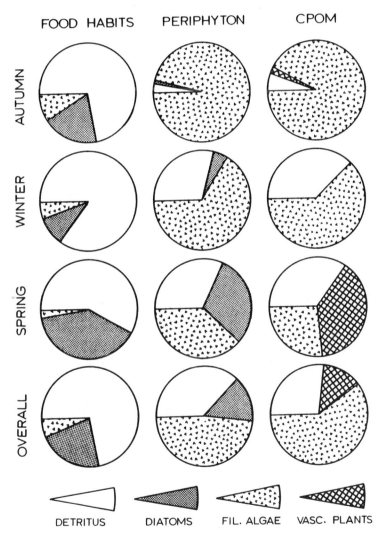

Figure 2.6. Mean percentage composition of major dietary components of herbivore–detritivores compared with the composition of these components in periphyton and coarse particulate organic matter (CPOM) at an altitude of 1896 m in Piceance Creek, Colorado. (From Gray and Ward, 1979.)

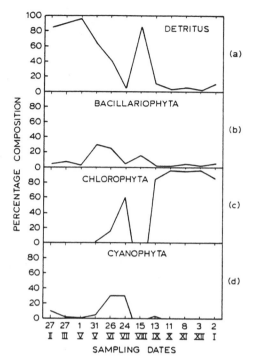

Figure 2.7.   Seasonal changes in the percentage composition of major epilithon components in a northern Colorado springbrook. (From Ward and Dufford, 1979.)

Diatoms (Bacillariophyta) at Site 2—*Achnanthes minutissima* dominant in fall; *Achnanthes* spp., *Cymbella* spp.,
and *Nitzschia amphibia* were predominant throughout the year. **Chlorophyta**—*Cladophora* sp. 90–99% of flora
throughout the year except December when *Oedogonium* sp. 90%. **Cyanophyta**—*Chamaesiphon incrustans* and
Gleotrichia sp.

A golden alga (Chrysophyceae), *Hydrurus foetidus*, occurs fairly frequently in the
South Gunnison drainage (Richardson and Gaufin, 1971). The presence of *Vaucheria*
sp. has also been indicated by its occurrence in insect guts (Pratt, 1938).

The diatoms (Bacillariophyceae) are the most diverse group of algae in the Gunnison. *Cymbella* sp. appears to be rare throughout, occurring sparingly on rocks in the
rapids. *Diatoma* sp. is one of the most abundant and widespread epilithic diatoms
found in the stream. *Gomphonema* sp. is found occasionally but does not appear to be
abundant. *Melosira* sp. and *Meridion* sp. are present but are not common. *Synedra* sp.
is extremely prolific, living on the surfaces of rocks and in the debris between rocks,
while *S. ulna* frequently occurs epiphytically on filamentous green algae. *Gomphonema ehrengregii*, *Fragilaria* sp., *Diatoma hiemale*, *Navicula* sp., and *Tabellaria* sp.
are also present (Pratt, 1938; Richardson and Gaufin, 1971).

In the Dolores River, flocculent detrital mats containing large numbers of *Cymbella* sp., *Fragilaria* sp., *Gomphonema* sp., *Navicula* sp., and *Synedra* sp. form over
the bottom in September. These mats disappear by late October, and the rocks support significant growths of *Cladophora* sp. (Fuller and Stewart, 1979).

In Piceance Creek, 31 and 29 species of diatoms are known from two sites. *Navicula viridula* is the most abundant at both sites. Other common species include *Navicula radiosa, N. rhynchocephala, Achnanthes lanceolata, A. minutissima,* and *Cocconeis pediculus.* The diatom communities peak in the winter at 2103 m altitude and in the spring at 1896 m, when runoff contributes to stream nutrification (Gray and Ward, 1979).

Eighty-eight species of diatoms, representative of 29 genera, are known from a hard Colorado springbrook (elevation 1597 m). Maximum relative abundances have been noted in May at four different lotic sites. An *Achnanthes/Cymbella* complex is abundant, with *Achnanthes minutissima* as the dominant *Achnanthes. Nitzschia amphibia* reaches maximum relative abundances in late March; however, it is not as dominant in the summer diatom flora. *Cocconeis placentula* displays a significant seasonal fluctuation, with greatest development during the winter (Ward and Dufford, 1979).

The red alga *Batrachospermum monoliforme* occurs in a northern Colorado springbrook (elevation 1597 m) (Ward and Dufford, 1979).

Bryophyta
Scleropodium obtusifolium occurs above 2316 m altitude and is abundant in the upper South Gunnison River. It forms dense mats on rock surfaces (Pratt, 1938).

Aquatic Vascular Plants
Potamogeton sp. is one of the characteristic plants of the lower Gunnison River. It forms extensive beds in backwaters of moderate depth. This plant requires a sandy or silty bottom to root. It occurs occasionally in fairly swift water if there is a soft bottom. During the late summer it serves as a refuge for young suckers (Pratt, 1938).

Protozoa
A few protozoans have been observed in the South Gunnison drainage. *Vorticella* sp. colonies occur under rocks, and several species of *Amoeba* inhabit the surfaces of dead leaves in marginal pools. *Paramecia* sp. is sometimes found to be common in colonies of *Spirogyra* sp., in the recessional pools. *Gregarina* sp. has been found in insect guts (Pratt, 1938).

Rotifera
Rotifers occur occasionally in marginal pools (Gunnison River). They are probably not endemic to the stream but are introduced into the pools from adjacent meadows (Pratt, 1938).

Mollusca
Ancylus sp. (Ancylidae), a herbivore–detritivore, and *Physa* sp. (Physidae), an omnivore, are common on stones in shallow shore rapids in the lower river, around 2316 m altitude. *Physa* sp. is also found in marginal pools (Gunnison River) (Pratt, 1938).

Pisidium casertanum (Sphaeriidae), an omnivore, occurs in the South Gunnison River and some of its tributaries, below 2300 m. It inhabits muddy bottoms in areas with little or no flow. It is often associated with *Potamogeton* sp. and filamentous algae (Wu, 1978).

Pisidium compressum (Sphaeridae), a detritivore, occurs rarely in the Gunnison drainage at approximately 2300 m altitude. It occurs in slightly turbid to clear water, at depths of 1 ft or less. It prefers a mud substratum and is often associated with *Potamogeton* sp. and algae (Wu, 1978).

Molluscs are rare if they occur in the upper river. *Physa* sp. appears to be the only mollusc that serves as a food source for trout in the Gunnison River (Pratt, 1938).

Pisidium sp., a detritivore; *Lymnaea* sp. (Lymnaeidae), an omnivore; and *Physa* are known to occur in a northern Colorado springbrook (elevation 1597 m) on most substrata. *Physa* sp. is the only snail in this stream found in moss (Ward and Dufford, 1979).

Platyhelminthes
Dugesia dorotocephala (Planariidae), a detritivore, occurs in a northern Colorado springbrook (elevation 1597 m) on all substrata, including moss (Ward and Dufford, 1979) (Table 2.4).

Annelida
Helodrilus sp. (Lumbricidae), a herbivore–detritivore, occurs in the gravel between rocks in swift rapids. This hyporheic species is usually found at least 10 in. below the surface of the gravel and is most common in the lower river (Pratt, 1938).

Eiseniella tetraedra (Lumbricidae), a herbivore–detritivore, occurs on all substrata except plant beds in a northern Colorado springbrook (elevation 1597 m). Permanent aquatic populations are apparently restricted to lotic reaches with reduced flow fluctuation (Ward and Dufford, 1979).

Pristina sp. (Naididae), a herbivore–detritivore, occurs in a northern Colorado springbrook (elevation 1597 m). It is the most abundant organism on sand (Ward and Dufford, 1979).

Three species of leeches (Hirudinea) occur in lotic habitats within the South Gunnison River drainage. *Erpobdella punctata*, a carnivore, is found throughout the entire state of Colorado at altitudes of 1000 to 2500 m and appears to be extremely adaptable. *Helobdella stagnalis*, a carnivore, is the most numerous and widely distributed leech in Colorado and ranges from 1000 to 3200 m in altitude; this species is common to most substrata except sand. *Glossiphonia complanata*, a carnivore, is also widespread and common throughout the state, being found at altitudes between 1000 and 2500 m, primarily under rocks. These three leeches are well adapted to wide ranges in total alkalinity concentrations. All three also serve as a food source for trout.

H. stagnalis is the only leech in the state found above 2500 m. This small, secretive, euryokous, and euryzonal leech is apparently more tolerant than larger species of turbulent mountain streams; it also seems to tolerate the greatest range of temperatures (5 to 21.5°C).

TABLE 2.4. *Species List: Colorado Springbrook, Foothills (Elevation 1597 m) of Rocky Mountains*

Taxon	V	S	M	G	C	Source[b]	Feeding Habits[c]
SUPERKINGDOM PROKARYOTAE							
KINGDOM MONERA							
Division Cyanophycota							
Class Cyanophyceae							
Order Nostocales							
Family Chamaesiphonaceae							
Entophysalis incrustans	×					WD	
Family Nostocaceae							
Anabaena sp.					×	WD	
Family Oscillatoriaceae							
Oscillatoria princeps					×	WD	
Phormidium sp.					×	WD	
SUPERKINGDOM EUKARYOTAE							
KINGDOM PLANTAE							
Subkingdom Thallobionta							
Division Chlorophycota							
Class Chlorophyceae							
Order Chlorococcales							
Family Characiaceae							
Characium sp.	×				×	WD	
Order Ulotrichales							
Family Chaetophoraceae							
Chaetophora sp.					×	WD	
Stigeoclonium sp.					×	WD	
Family Microsporaceae							
Microspora sp.					×	WD	
Family Ulotrichasceae							
Ulothrix zonata					×	WD	
Order Zygnematales							
Family Zygnemataceae							
Spirogyra sp.					×	WD	
Zygnema sp.					×	WD	
Order Cladophorales							
Family Cladophoraceae							
Cladophora glomerata	×				×	WD	
Order Oedogoniales							
Family Oedogoniaceae							
Oedogonium sp.					×	WD	
Class Bacillariophyceae							
Order Achnanthales							
Family Achnanthaceae							
Achnanthes minutissima					×	WD	
Cocconeis placentula					×	WD	
Order Bacillariales							
Family Nitzschiaceae							
Nitzschia amphibia					×	WD	
Division Euglenophyta							
Class Euglenophyceae							
Order Euglenales							
Family Euglenaceae							
Euglena sp.	×	×			×	WD	

(continues)

TABLE 2.4.   *Continued*

Taxon	Substrate[a]					Source[b]	Feeding Habits[c]
	V	S	M	G	C		
Phacus sp.	×		×		×	WD	
Division Rhodophyta							
Class Rhodophyceae							
Order Nemalionales							
Family Batrachospermaceae							
Batrachospermum monoliforme					×	WD	
KINGDOM ANIMALIA							
Subkingdom Protozoa							
Phylum Platyhelminthes							
Class Turbellaria							
Order Tricladida							
Family Planariidae							
Dugesia dorotocephala	×	×	×	×	×	WD	D
Phylum Annelida							
Class Oligochaeta							
Order Lumbriculida							
Family Lumbricidae							
Eiseniella tetraedra		×	×	×	×	WD	HD
Order Haplotaxida							
Family Naididae							
Pristina sp.		×	×			WD	HD
Phylum Arthropoda							
Subphylum Mandibulata							
Class Crustacea							
Subclass Malacostraca							
Order Amphipoda							
Family Talitridae							
Hyallela azteca						WD	O
Class Insecta							
Order Coleoptera							
Family Elmidae							
Optioservus sp.	×				×	WD	O
Family Simuliidae							
Simulium hunteri					×	WD	O
S. latipes					×	WD	O
Order Ephemeroptera							
Family Baetidae							
Baetis tricaudatus					×	WD	DO
Baetis spp.					×	WD	DO
Callibaetis nigritus					×	WD	O
Family Ephemerellidae							
Ephemerella inermis					×	WD	O
Family Heptageniidae							
Heptagenia spp.		×			×	WD	O
Family Tricorythidae							
Tricorythodes sp.					×	WD	O
Order Trichoptera							
Family Hydropsychidae							
Cheumatopsyche sp.					×	WD	O
Hydropsyche spp.					×	WD	O
Family Limnephilidae							
Hesperophylax incisus	×		×	×	×	WD	HD
Phylum Mollusca							

TABLE 2.4. *Continued*

Taxon	V	S	M	G	C	Source[b]	Feeding Habits[c]
			Substrate[a]				
Class Gastropoda							
Order Basommatophora							
Family Lymnaeidae							
Lymnaea spp.		×	×	×	×	WD	O
Family Physidae							
Physa sp.	×	×	×	×	×	WD	O
Class Pelecypoda							
Order Heterodonta							
Family Sphaeriidae							
Pisidium sp.	×	×	×		×	WD	D

[a]V, vegetation and debris; S, sand; M, mud; G, gravel; C, cobble/rubble.
[b]WD, Ward and Dufford (1979).
[c]Feeding habits: D = detritivore; DO = detritivore–omnivore; HD = herbivore–detritivore; O = omnivore.

G. complanata and *H. stagnalis* often occur together. The food habits of these two leeches are similar, yet subtle dietary differences are present. Both feed on chironomid larvae, oligochaetes, and small snails. The snails, however, are the first preference of *G. complanata*, while *H. stagnalis* prefers oligochaetes and chironomid larvae.

High-Altitude Streams. In high-altitude Colorado mountain streams and rivers, a low dissolved salt content combines with low temperatures and rapid currents, to provide a harsh environment. Only one lotic species, the leech *H. stagnalis*, seems capable of tolerating these conditions. Low dissolved salt content probably prevents the survival of food organisms preferred by other species; however, the effects of prolonged low temperature (<11°C) on reproduction, and the effects of rapid current on attachment, are apparently the most important limiting factors since many aquatic insects are available as food above 2500 m (Herrmann, 1970).

Crustacea

Gammarus sp. (Gammaridae), an omnivore, occurs rarely in the South Gunnison River. It is known only from pooled areas and may be an invading species (Pratt, 1938).

Hyallela azteca (Talitridae), an omnivore, occurs in a northern Colorado spring-brook (elevation 1597 m) on all substrata except moss (Ward and Dufford, 1979).

Insecta

Plecoptera. Plecoptera distribution is unquestionably related to altitude. Figure 2.8 shows the altitudinal distributions of South Gunnison stoneflies. There are three groups of stoneflies in the Gunnison drainage with respect to altitudinal distribution (Knight, 1965): (1) stenotherms, limited to low altitudes; (2) stenotherms, limited to high altitudes; and (3) eurytherms, occurring over a wide altitudinal range.

The effects of altitude on stonefly distribution are chiefly the effects of temperature and, more indirectly, food supply and stream size. Temperature is the primary limiting factor for most stoneflies and thus greatly influences Plecoptera distribution.

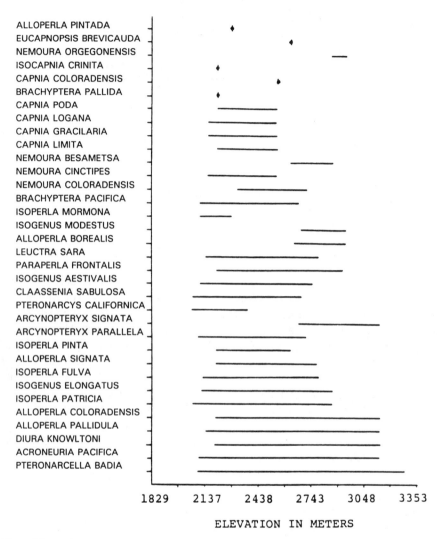

Figure 2.8.   Altitudinal range of stoneflies from 1829 to 3353 m in the South Gunnison River drainage. (From Knight and Gaufin, 1966.)

Many stoneflies can tolerate only minimal fluctuations in temperature and therefore are restricted to very narrow distributional ranges. For successful reproduction, stenotherms are most frequently restricted by temperature requirements.

Many stoneflies, already limited to certain ranges by temperature, may have developed specific feeding structures and functions to more efficiently utilize the available food resources at particular altitudes (Knight, 1965).

There is a significant correlation between altitude and population dynamics of two groups of stoneflies (Figure 2.9). Stonefly nymphs of the suborder Filapalpia are

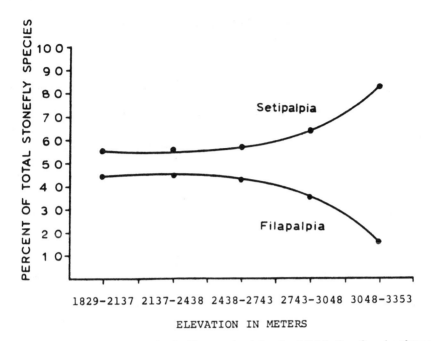

Figure 2.9.   Relationship between suborder (Plecoptera) and elevational distribution of species of stone-flies in the Gunnison River drainage. (From Knight, 1965.)

generally herbivorous, while nymphs of the suborder Setipalpia are usually carnivorous. The data indicate that the numbers of Setipalpia species are common at all altitudes and increase with altitude. This increased abundance of carnivorous stoneflies at high altitudes is no doubt related to the relative paucity of algae and other vegetation in these regions (Knight, 1965).

Current is also an important factor in stonefly distribution since it controls dissolved oxygen concentrations, food and microhabitat availability, and substrata composition. Stonefly habitats can be divided into six types (Knight and Gaufin, 1967):

1. *Temporary streams or snow flood streams:* cascades and trickles that arise from melting snows and for a time may carry considerable quantities of water, which diminish by summer; substrate ranges from 2.5 to 10 cm in diameter.
2. *High small stony streams:* small rapid streams found in mountainous regions and ranging from 0.3 to 4.5 m in width; may carry enormous masses of water for brief periods in spring; substrate ranges from 2.5 to 25.4 cm in diameter; altitude over 2440 m.
3. *Low small stony streams:* similar to high small stony streams; altitude below 2440 m.
4. *Stony rivers:* 9 to 27 m wide with a stony substrate (2.5 to 61.5 cm in diameter) and a rapid current; carries enormous quantities of water in the spring.

5. *Constant rivers:* 27 to 55 m wide; carries a large volume of water that is only slightly diminished in summer; substrate ranging from sand to boulders 1.2 m in diameter.
6. *Sluggish rivers:* large or small with widespread areas of muddy bottom and rooted macrophytes; substrate ranges from sand to rock 61.5 cm in diameter.

Gunnison stoneflies appear to select for stream type. Table 2.5 shows the associations of stonefly species with various stream types. It seems apparent from these data that stoneflies occur more frequently in streams with high concentrations of dissolved oxygen and minimal water temperature fluctuations.

TABLE 2.5. *Stream Types of the Species of Stoneflies*

Stonefly Species	Small Stony Streams			Stony Rivers	Constant Rivers	Sluggish Rivers
	Temporary Streams	High Streams	Lower Streams			
Acroneuria pacifica	×	×	×	×	×	×
Alloperla borealis	×			×		
A. coloradensis		×	×	×	×	
A. pallidula	×	×	×	×	×	
A. pintada			×			
A. signata	×		×	×	×	×
Arcynopteryx parallela			×	×	×	×
A. signata	×	×		×		
Brachyptera pacifica				×	×	
B. pallida					×	
Capnia coloradensis				×		
C. gracilaria			×	×	×	
C. limita				×	×	
C. logana			×	×	×	
C. poda			×	×		
Claassenia sabulosa	×		×	×	×	×
Diura knowltoni		×	×	×	×	
Eucapnopsis brevicauda					×	
Isocapnia crinita					×	
Isogenus aestivalis			×	×	×	×
I. elongatus	×		×	×	×	×
I. modestus				×	×	
Isoperla fulva	×	×	×	×	×	×
I. morona			×	×		
I. patricia	×		×	×	×	×
I. pinta				×	×	×
Leuctra sara		×	×	×	×	
Nemoura besametsa				×	×	
N. cinctipes			×	×		
N. coloradensis	×		×	×		
N. oregonensis				×		
Paraperla frontalis				×		
Pteronarcella badia		×	×	×	×	×
Pteronarcys californica			×	×	×	×

Source: After Knight and Gaufin (1967).

The eight species collected from high-altitude small stony streams are apparently more tolerant of very low temperatures and have life cycles that can be completed during the short ice-free season. Low-altitude small stony streams, stony rivers, and constant rivers have very diverse stonefly communities because of the longer ice-free seasons and the greater diversity of habitat and food.

Several large species requiring more than one year for development are excluded from small streams that become greatly reduced in size in late summer; *Nemoura oregonensis* (detritivore), *Capnia coloradensis* (detritivore), and *Paraperla frontalis* (omnivore) are unique to the stony rivers, which are large enough to provide sufficient habitat, yet still have sufficient current to provide high dissolved oxygen concentrations. Medium-sized streams often have the greatest variety of stoneflies, because they provide resources for both large and small species (Figure 2.10) (Knight and Gaufin, 1967).

Species diversity is relatively low in sluggish rivers, primarily because of the lower concentrations of dissolved oxygen (Knight and Gaufin, 1967). Local associations of stonefly species have also been correlated (Figure 2.11). These data do not take into account the abundance of each species at a site, only its presence (Knight and Gaufin, 1967).

Another area of ecological research on Plecoptera is diet and feeding habits. This order contains the largest carnivorous (*Acroneuria* spp.) and the largest herbivorous (*Pteronarcys* sp.) insects found in the Gunnison (Pratt, 1938).

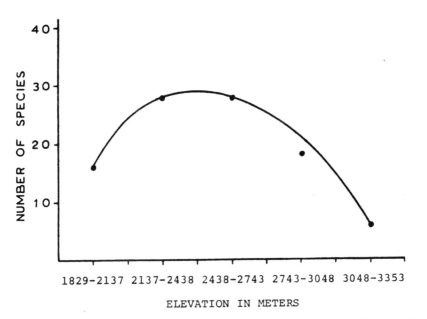

Figure 2.10. Relationship between elevation and number of stonefly species in the Gunnison River drainage. (From Knight, 1965.)

Brachyptera fallida	Arcynopteryx signata	Arcynopteryx parallela	Isogenus aestivalis	Isogenus elongatus	Isogenus modestus	Diura knowltoni	Isoperla fulva	Isoperla hormona	Isoperla patricia	Isoperla pinta	Alloperla pintada	Alloperla pallidula	Alloperla borealis	Alloperla coloradensis	Alloperla signata	Paraperla frontalis	Acroneuria pacifica	Claassenia sabulosa	Capnia poda	Capnia logana	Capnia coloradensis	Capnia gracilaria	Capnia limita	Isocapnia orinita	Nemoura besametsa	Nemoura cinctipes	Nemoura oregonensis	Nemoura coloradensis	Eucapnopsis brevicauda	Leuctra sara	Brachyptera pacifica	Pteronarcys californica	
1.9	6.5	32.7	28.3	31.6	1.9	10.7	41.3	2.0	58.5	9.4	1.9	14.3	0.0	1.7	24.1	3.8	53.0	41.0	10.7	11.1	1.9	5.7	7.4	1.9	3.5	5.7	0.0	3.8	0.0	12.7	5.6	25.1	Pteronarcella badia
	0.0	4.5	4.0	4.4	0.0	0.0	3.3	0.0	2.5	20.0	0.0	0.0	0.0	5.5	5.0	0.0	2.0	2.9	8.3	14.3	0.0	0.0	1.7	100.0	0.0	0.0	0.0	0.0	0.0	0.0	25.0	5.6	Brachyptera pallida
		2.9	8.3	8.8	21.4	26.3	7.3	0.0	1.9	0.0	0.0	8.3	13.3	14.3	6.3	6.3	5.0	2.1	0.0	0.0	0.0	0.0	0.0	0.0	23.5	0.0	14.3	0.0	0.0	14.3	5.9	0.0	Arcynopteryx signata
			35.3	37.5	4.3	10.3	34.2	4.5	32.6	13.6	0.0	6.5	0.0	21.9	24.2	4.3	27.3	41.0	8.0	0.0	0.0	4.2	7.7	4.8	0.0	0.0	0.0	4.3	0.0	10.7	13.6	30.0	Arcynopteryx parallela
				33.3	3.7	16.7	34.1	3.8	47.7	14.8	0.0	19.4	0.0	16.2	31.4	3.7	27.6	46.3	15.2	6.7	0.0	0.0	6.7	4.0	6.7	0.0	0.0	0.0	0.0	12.9	11.5	26.5	Isogenus aestivalis
					0.0	22.2	35.9	0.0	34.0	16.0	0.0	6.1	0.0	17.1	26.5	8.3	24.1	23.4	33.3	11.1	4.3	8.0	16.0	4.3	15.4	3.8	4.2	4.0	0.0	3.1	12.5	24.2	Isogenus elongatus
						8.3	3.1	0.0	0.0	0.0	0.0	7.1	20.0	5.0	0.0	20.0	2.0	2.8	0.0	0.0	0.0	0.0	0.0	0.0	0.0	0.0	25.0	0.0	0.0	8.3	0.0	0.0	Isogenus modestus
							17.6	0.0	6.4	6.7	0.0	10.0	0.0	16.7	15.4	8.3	9.3	10.0	9.5	5.9	0.0	0.0	0.0	0.0	13.3	0.0	0.0	0.0	0.0	0.0	0.0	7.7	Diura knowltoni
								3.2	27.3	12.5	0.0	13.5	0.0	37.1	28.2	3.1	41.1	25.0	13.2	15.2	3.3	6.3	8.8	3.3	12.1	6.3	0.0	3.1	3.3	14.3	9.7	14.3	Isoperla fulva
									2.4	0.0	0.0	7.7	0.0	5.3	4.8	0.0	4.1	2.9	0.0	0.0	0.0	0.0	0.0	0.0	0.0	0.0	0.0	0.0	0.0	9.1	0.0	5.3	Isoperla normona
										9.5	0.0	15.6	0.0	9.4	25.0	4.9	36.9	51.0	20.5	11.6	2.5	2.3	9.3	2.5	2.2	2.3	2.4	2.4	0.0	4.2	7.3	28.9	Isoperla patricia
											0.0	0.0	0.0	0.0	0.0	0.0	2.0	0.0	0.0	0.0	0.0	0.0	0.0	0.0	0.0	25.0	0.0	0.0	0.0	0.0	0.0	0.0	Alloperla pintada
												5.9	0.0	9.1	8.3	12.5	5.8	14.3	26.7	7.7	0.0	11.1	30.0	16.7	0.0	0.0	0.0	0.0	0.0	0.0	11.1	14.3	Isoperla pinta
													7.1	15.4	10.3	14.3	15.1	21.1	8.7	11.1	0.0	6.7	0.0	0.0	0.0	6.7	7.7	0.0	0.0	10.0	0.0	11.1	Alloperla pallidula
														0.0	0.0	20.0	2.0	0.0	0.0	0.0	0.0	0.0	0.0	0.0	0.0	0.0	25.0	0.0	0.0	0.0	0.0	0.0	Alloperla borealis
															31.0	5.0	34.1	13.3	10.7	4.2	0.0	0.0	8.7	5.6	19.0	0.0	0.0	10.5	5.6	33.3	4.8	9.1	Alloperla coloradensis
																4.5	21.1	23.3	10.0	7.7	0.0	0.0	8.0	5.0	8.0	0.0	0.0	4.5	0.0	20.0	9.1	18.8	Alloperla signata
																	4.0	2.8	14.3	10.0	33.3	16.7	11.1	0.0	0.0	16.7	25.0	0.0	0.0	0.0	0.0	0.0	Paraperla frontalis
																		38.3	17.0	11.8	2.0	6.0	7.7	2.0	5.7	6.0	0.0	2.0	2.0	15.4	6.0	21.8	Acroneuria pacifica
																			17.5	7.7	0.0	2.9	8.1	2.9	0.0	0.0	0.0	0.0	0.0	10.0	5.6	40.5	Claassenia sabulosa
																				50.0	7.7	21.4	53.8	7.7	0.0	13.3	0.0	0.0	0.0	0.0	13.3	14.8	Capnia poda
																					12.5	33.3	18.2	12.5	0.0	33.3	0.0	0.0	0.0	0.0	33.3	4.0	Capnia logana
																						25.0	14.3	0.0	0.0	25.0	0.0	0.0	0.0	0.0	0.0	0.0	Capnia coloradensis
																							22.2	0.0	0.0	60.0	0.0	0.0	0.0	0.0	0.0	4.8	Capnia gracilaria
																								14.3	0.0	10.0	0.0	0.0	0.0	0.0	10.0	13.6	Capnia limita
																									0.0	0.0	0.0	0.0	0.0	0.0	25.0	5.6	Isocapnia crinita
																										0.0	12.5	0.0	14.3	13.3	10.0	0.0	Nemoura besametsa
																											0.0	0.0	0.0	0.0	0.0	0.0	Nemoura cinctipes
																												0.0	0.0	0.0	0.0	0.0	Nemoura oregonensis
																													0.0	8.3	0.0	0.0	Nemoura coloradensis
																														0.0	0.0	0.0	Eucapnopsis brevicauda
																															0.0	12.0	Lectra sara
																																10.0	Brachyptera pacifica

Figure 2.11. Association diagram of the stoneflies in the Gunnison River drainage. (From Knight and Gaufin, 1967.)

The stoneflies can be logically divided into three categories on the basis of feeding habits: (1) phytophagous, (2) polyphagous or omnivorous, and (3) predominantly carnivorous. Some species overlap categories because of the need to modify their diet in response to environmental conditions (Richardson and Gaufin, 1971).

Most Plecoptera are polyphagous with a propensity toward either herbivory or carnivory. *Pteronarcys californica*, an omnivore–carnivore, and *Pteronarcella badia*, an omnivore–carnivore, are stoneflies with diets that consist mostly of plant matter. *Arcynopteryx signata* (carnivore), *A. parallela* (carnivore), *Isoperla fulva* (carnivore), and *Isogenus elongatus* (carnivore) on the other hand, lean more toward carnivory. Because of omnivorous dietary habits, and hence the ability to utilize available plant and animal matter, these species generally possess wider distributional ranges.

Setipalpian stoneflies are primarily carnivorous. Their diets change during development, from chironomids and diatoms in the earliest instars, to primarily chironomids, to a broader diet in which mayflies increase proportionately. As the predators grow, the diversity of their diet increases.

At a given site, setipalpian stonefly diets appear similar for individuals of a given size, regardless of species. Diet differs, however, between sites for individuals of a given species and size class, probably because of prey availability. There are also differences in the diets of individuals of the same species at the same site but of different size classes. It appears that setipalpian stonefly feeding habits are influenced primarily by the size of the individual, not by species (Allan, 1982a).

There are also certain differences in the altitudinal distribution of these setipalpian stoneflies; this could be a result of competition (Table 2.3). In areas of overlap, competition may be minimized by microhabitat specialization or by body size differences, resulting in selection of different prey. Alternatively, co-occurrence could be the result of prey abundance and the absence of competition, or simply the result of stream drift.

Pteronarcella badia (Pteronarcidae), an omnivore–carnivore, occurs in stony rivers, constant rivers, high-altitude small stony streams, low-altitude stony streams, and sluggish rivers. This species is a eurytherm and ranges from 2101 to 3283 m altitude. It is found in relatively quiet areas and is especially abundant in leaf masses and other accumulations of debris. *P. badia* occurs with *Pteronarcys californica* 25.1% of the time (Knight, 1965; Knight and Gaufin, 1966; Pratt, 1938; Richardson and Gaufin, 1971).

Pteronarcella badia is apparently an omnivore–carnivore that alters feeding habits with the availability of food. This species has a herbivorous diet at low altitudes and a carnivorous diet at higher altitudes. Food items usually consist of vascular plant tissue, mosses, diatoms, and filamentous green algae. In areas where vegetation is scarce, *P. badia* ingests Ephemeroptera, Plecoptera, and chironomid larvae. This shift in feeding habits may have developed as an energy-efficient method of obtaining and/or utilizing food (Knight and Gaufin, 1966; Richardson and Gaufin, 1971).

In the Dolores River, *Pteronarcella badia* nymphs (omnivore–carnivores) inhabit debris in slowly moving water. Gut contents of early instars in September contain components of the detritus–diatom mats abundant at this time. After mat disintegration the diet is still predominantly detritus. Later instars shift to a more polyphagous diet that includes moss (Fuller and Stewart, 1979).

Pteronarcys californica (Pteronarcidae), an omnivore–carnivore, is the largest and most conspicuous of all the Gunnison stoneflies. It is a stenotherm limited to low altitudes and inhabits moderate rapids, between 2060 and 2377 m altitude. It is the most numerous stonefly species at all sites at which it is found. Nymphs are primarily nocturnal. Their abundance, large size, and roving behavior make them a favorite food of the trout (Knight, 1965; Knight and Gaufin, 1966, 1967).

Pteronarcys californica nymphs (omnivore–carnivores) occur in low-altitude small stony streams, stony rivers, constant rivers, and sluggish rivers. This species is probably not a permanent resident in the low-altitude small stony streams but a transient from the large rivers. Nymphs occur primarily among rocks and rubble, where there is an accumulation of trash or luxuriant algal growth. The diet of *Pteronarcys californica* is composed of detritus (79.7%), filamentous green algae (6.6%), and accidentally ingested epiphytic diatoms (2.1%). Another 8% is composed of chironomid and Trichoptera larvae and Ephemeroptera nymphs (especially the baetids and *Ephemerella* sp., an omnivore). Some changes in diet are due to seasonal abundances of food. The herbivorous diet and a three-year life cycle may explain why this species is never found at high elevations, where vegetation is sparse and the ice-free season is of short duration. *P. californica* (omnivore–carnivore) is found with *Claassenia sabulosa* (carnivore) 40.5% of the time (Knight, 1965).

Pteronarcys californica, an omnivore–carnivore, inhabits the swiftest areas of the Dolores River. Large amounts of detritus are eaten in all months. *Cladophora* sp. comprises 40% of the diet in October. Moss is more commonly consumed in the spring (Fuller and Stewart, 1979).

Skwala parallela[1] (Perlodidae), a carnivore, occurs at 2316 m altitude. It is a small species and appears to be quite rare (Pratt, 1938). Nymphs of *Arcynopteryx signata* (Perlodidae), a carnivore, occur in temporary streams, high-altitude small stony streams, and stony rivers. They are stenothermic and limited to high altitudes between 2332 and 3139 m. Nymphs are usually found in the swiftest, most aerated portion of the stream, clinging to the undersides of rocks and rubble. This species is found with *Nemoura besametsa*, a detritivore, 23.5% of the time (Knight, 1965; Knight and Gaufin, 1966, 1967).

A. signata, a carnivore, is primarily a carnivore, while also consuming some detritus. Baetid and heptageniid ephemeropterans and chironomid and simuliid dipteran larvae constitute the major portion of its diet. The variety in the *A. signata* diet indicates that this species feeds randomly on macroinvertebrates. Vegetation may be ingested incidentally or in the absence of prey (Richardson and Gaufin, 1971).

Arcynopteryx parallela (Perlodidae) nymphs (carnivores) occur in low-altitude small stony streams, stony rivers, constant rivers, and sluggish rivers. Altitude ranges from 2101 to 2716 m. This species is found in association with *Isogenus elongatus*, a carnivore, 37.5% of the time (Knight, 1965).

A. parallela occurs beneath rocks in swift water. This species is an omnivore, with a diet composed (58.7%) of animal matter (primarily Ephemeroptera nymphs and

[1]Pratt (1938) called this species *Perlodes americana.*

chironomid larvae). Detritus is an important part of the *A. parallela* diet; filamentous green algae and diatoms are also eaten (Richardson and Gaufin, 1971).

Isogenus aestivalis[2] (Perlodidae) nymphs and adults (carnivores) are found in or along low-altitude small stony streams, stony streams, constant rivers, and sluggish rivers. They are eurytherms ranging from 2118 to 2751 m altitude. This species is found with *Isogenus elongatus* (carnivore) 33.3% of the time and *I. modestus* (carnivore) 3.7% of the time. *I. aestivalis* (carnivore) is commonly found crawling on stones in fast riffle areas. It is a typical carnivore, feeding on smaller animals as well as members of its own species (Knight, 1965). Adults emerge between June 15 and July 1 under small rocks near the shore (Knight and Gaufin, 1966; Pratt, 1938).

Isogenus elongatus (Perlodidae) nymphs (carnivores) occur in temporary streams, low-altitude small stony streams, stony rivers, constant rivers, and sluggish rivers. They are eurythermic ranging from 2118 to 2865 m. This species is never found with *I. modestus*. Nymphs are commonly found clinging to the undersides of rocks in fast riffle areas (Knight, 1965; Knight and Gaufin, 1966, 1967).

The *I. elongatus* diet consists primarily of animal matter (64.9%), but under certain conditions it feeds almost exclusively on detritus. This species seems to prefer smaller food items, such as simuliid and chironomid larvae. The vegetative portion of its diet is primarily filamentous green algae and diatoms (Richardson and Gaufin, 1971).

Isogenus modestus (Perlodidae) nymphs (carnivores) occur in stony rivers and constant rivers. They are high-altitude stenotherms, ranging from 2682 to 2946 m altitude. This species is found clinging to the undersides of rocks in fast riffle areas. It is mainly carnivorous, feeding on smaller animals (Knight, 1965; Knight and Gaufin, 1966).

Diura knowltoni (Perlodidae) nymphs (carnivores) occur in high-altitude small stony streams, low-altitude small stony streams, stony rivers, and constant rivers. Nymphs inhabit fast riffle areas, where they cling to the undersides of rocks or rubble. They are eurythermic, ranging from 2201 to 3139 m. This species occurs with *Arcynopteryx signata*, a carnivore, 26.3% of the time (Knight, 1965; Knight and Gaufin, 1966).

Isoperla fulva (Perlodidae) nymphs and adults (carnivores) occur in temporary streams, high-altitude small stony streams, low-altitude small stony streams, stony rivers, constant rivers, and sluggish rivers. Nymphs are found on a variety of substrata. They are commonly taken from fast currents, in either shallow or deep water, where a suitable substratum is present. They are eurytherms occurring from 2118 to 2786 m. They occur with *Acroneuria pacifica*, a carnivore, 41% of the time, and with *Isoperla* spp. less frequently (Knight, 1965; Knight and Gaufin, 1966).

I. fulva, a carnivore, is found beneath stones and other materials in swift streams. Its diet is 70.4% animal matter. Ephemeroptera nymphs and chironomid larvae are commonly ingested, while terrestrial arthropods, Plecoptera nymphs, Trichoptera, and simuliid larvae are consumed in smaller proportions. Filamentous green algae are the most commonly consumed vegetation, while diatoms are rarely utilized as a food source. Detritus is a very important part of the *I. fulva* diet (Richardson and Gaufin, 1971).

---

[2] Pratt (1938) called this species *Perla aestivalis*.

In the Dolores River, *Isoperla fulva* feeds almost exclusively on chironomids and simuliids (Fuller and Stewart, 1979).

Isoperla modesta gr. sp. (Perlodidae), a carnivore, occurs between 2316 and 2408 m altitude (Pratt, 1938).

Isoperla patricia (Perlodidae) nymphs and adults (carnivores) occur in or along temporary streams, low-altitude small stony streams, stony rivers, constant rivers, and sluggish rivers. The nymphs are found in swift, well-aerated water, clinging to the undersides of stones and other objects. They are eurytherms occurring from 2042 to 2865 m. This species is most commonly found with *Pteronarcella badia*, an omnivore–carnivore (Knight, 1965; Knight and Gaufin, 1966).

Isoperla patricia, a carnivore, is the most common invertebrate predator in Piceance Creek. At 1896 m the only animals consumed are chironomids and *Baetis* sp. At 2103 m, chironomids, *Baetis*, *Capnia* sp., and oligochaetes comprise the diet. Diet composition varies with the size class of *I. patricia* nymphs. *Capnia* and *Baetis* are the typical prey of large nymphs, while small chironomids are the preferred prey of small nymphs. The extended hatching period and food resources partitioning apparently reduce intraspecific competition (Gray and Ward, 1979).

Isoperla pinta (Perlodidae) nymphs (carnivores) are found in stony rivers, constant rivers, and sluggish rivers. It is a low-altitude stenotherm most commonly found in swift riffle areas, between 2195 and 2621 m altitude. This species is associated with *Capnia poda* 26.7% of the time (Knight, 1965; Knight and Gaufin, 1966).

Isoperla quinquepunctata[3] (Perlodidae) is a small carnivorous species that occurs in quiet water, between 2316 and 2408 m altitude. Adults emerge between June 15 and July 1 under small rocks along the shore (Pratt, 1938).

Megarcys signata (Perlodidae), a carnivore, is present from 2900 to 3350 m altitude and most common at the higher end of this range. Early instar nymphs are present by early June, and final instar nymphs disappear by early July. Growth is very rapid through the summer, until mid-October, when water temperature decreases. The maximum size is 12 to 17 mg for males and 25 to 40 mg for females. Chironomids comprise at least 75% of the animal prey consumed. An increase in prey diversity occurs as the size of the individual increases. *M. signata* sometimes consumes substantial amounts of diatoms (Allan, 1982).

Kogotus modestus (Perlodidae), a carnivore, occurs at altitudes of 2600 to 3350 m and is more common in the lower reaches of this range. Final instars are present through August, and maximum size is 2 to 5 mg. Chironomids are the most important element of the *K. modestus* diet (Allan, 1982a).

In the Dolores River, *Cultus aestivalis*, a carnivore, feeds almost exclusively on chironomids and simuliids (Fuller and Stewart, 1979).

Paraperla frontalis (Chloroperlidae) nymphs and adults (omnivores) are found in stony rivers. They are eurytherms and range from 2195 to 2946 m altitude. Nymphs occur in rapid riffle areas, where they are found under and among rocks. This species occurs with *Capnia coloradensis* 33.3% of the time (Knight, 1965; Knight and Gaufin, 1966).

---

[3] Pratt (1938) called this species *Isoperla 5-punctata*.

The Chloroperlidae *Sweltsa* sp., *Suwallia* sp., and *Triznaka* sp., omnivores, are exclusively carnivorous in May, feeding primarily on chironomid and simuliid larvae. During the summer there is a shift to a polyphagous diet, with the ingestion of large amounts of detritus (Fuller and Stewart, 1979).

Alloperla signata (Chloroperlinae), an omnivore–carnivore, occurs between 2195 and 2789 m altitude. It is small and appears to be rare. This species is found in deep holes where collection is very difficult (Pratt, 1938).

Acroneuria pacifica (Perlidae), a carnivore, is a large eurythermic stonefly occurring between 2073 and 3124 m altitude. It is common and quite abundant. These insects generally live among rocks in very swift water. Because of their flattened bodies and strong legs, they are ideally suited for survival in swift currents; they can flatten into the boundary layer and move rapidly when movement is necessary. They have a relatively long period of emergence, which lasts from July 1 until August 12. The nymphs leave the water, cling to the surface of a flat rock, and emerge. Adults are apparently nocturnal, remaining under rocks during the day (Knight and Gaufin, 1966, 1967; Pratt, 1938).

Male and female nymphs appear to be equally abundant. In adult collections, however, males are 15 times more numerous than females. This may be because male *Acroneuria* have vestigial wings, and only females of the species can fly. Low numbers of adult females could be attributed to their vulnerability when flying or to their ability to escape collection more easily (Pratt, 1938).

Acroneuria pacifica nymphs and adults (carnivores) occur in temporary streams, high-altitude small stony streams, low-altitude small stony streams, stony rivers, constant rivers, and sluggish rivers. Altitude ranges from 2060 to 3139 m. This species is associated with *Pteronarcella badia*, an omnivore–carnivore, 53% of the time. Nymphs are found in fast riffles clinging to the undersides of rocks and rubble (Knight, 1965).

A. pacifica is carnivorous, with 88.3% of its diet composed of animal matter. Ephemeropterans, especially *Epeorus longimanus*, *Ephemerella* spp., and baetids, are most commonly consumed. Chironomid larvae, Trichoptera, Plecoptera, Oligochaeta, Crustacea, and terrestrial Arthropoda are also ingested. This varied diet suggests random voracious carnivory. Vegetation is consumed only accidentally (Richardson and Gaufin, 1971).

Claassenia depressa[4] (Perlidae), a carnivore, is another large stonefly commonly found between 2256 and 2408 m altitude (Pratt, 1938).

Claassenia sabulosa[5] (Perlidae) nymphs (carnivores) occur in temporary streams, low-altitude small stony streams, stony rivers, constant rivers, and sluggish rivers. They are eurytherms ranging from 2060 to 3050 m altitude, although they are rare in the upper reaches of this range. This species is found with *Isoperla patricia*, a carnivore, 51% of the time. *C. sabulosa* nymphs (carnivores) occur under rocks in very rapid water, usually at moderate depths. Maximum size is 25 to 90 mg (Allan, 1982a; Knight, 1965; Knight and Gaufin, 1966, 1967; Pratt, 1938).

[4] Pratt (1938) called this species *Acroneuria depressa*.
[5] Pratt (1938) called this species *Perla languida*.

Diet is typically setipalpian, consisting chiefly of chironomids when small and an increase in prey diversity as size increases. Its diet is 91.4% animal matter. Its well-developed mouthparts and agility make it well suited for preying upon most macroinvertebrates. Although a variety of organisms are consumed, Ephemeroptera nymphs (especially baetids, heptageniids, and *Ephemerella* spp.), simuliid, chironomid, and Trichoptera larvae and Plecoptera nymphs are the most common. Small amounts of detritus and filamentous green algae are also eaten (Richardson and Gaufin, 1971).

In the Dolores River, *Claassenia sabulosa*, a carnivore, inhabits large rock areas of riffles. It is opportunistic, feeding on a variety of macroinvertebrates. The food habits of this species vary significantly with season, river system, and size class (Fuller and Stewart, 1979).

Brachyptera sp. (Nemouridae) is exclusively herbivorous, as indicated by habitat and morphology. This species resides beneath stones during the day, consuming accumulated detritus; at night it uses its specialized mouthparts to scrape algae off the upper surfaces of rocks. Filamentous green algae are the preferred food items (Richardson and Gaufin, 1971).

Zapada haysi (Nemouridae), a detritivore, is known to occur at 3000 and 3100 m altitude (Allan, 1981).

Altitudinal distributions for additional South Gunnison stoneflies are given in Figure 2.8.

In another Colorado hardwater stream, Trout Creek, the Plecoptera account for about 7% of the macroinvertebrate community. Although 10 species have been collected, the dominant stoneflies are *Triznaka signata* (omnivore) and *Prostoia besametsa* (detritivore) (Canton and Ward, 1981).

Ephemeroptera. Certain mayflies (e.g., *Baetis bicaudatus* (a detritivore–omnivore), *Rhithrogena* sp. (a detritivore–omnivore), *Cinygmula* sp. (an omnivore), *Epeorus longimanus* (an omnivore), and Ephemerellidae) are dominant food items in the diet of local trout. This phenomenon, however, is probably a result of the overwhelming abundance of Ephemeroptera in the stream and not because they are selected for. Trout predation has a notable effect on abundances of nymphs and larvae. Above the upper altitudinal limit of the trout (3250 m), there is a significant increase in invertebrate populations (Figure 2.12). The predator-free headwaters probably function as a source of downstream mayfly colonization and of trout prey (Allan, 1975, 1978b, 1981).

Mayflies are dominant in the stream invertebrate assemblage and are very rapid initial colonizers of virgin substrata. The peak of mayfly transformation occurs during late June and early July. At this time, thousands of *Baetis* sp. (detritivore–omnivores) and *Ephemerella* sp. nymphal exuviae and subimagoes (omnivores) can be observed floating on the surface of the water (Pratt, 1938). Ephemeroptera nymphs account for 62.5% of total invertebrate drift while accounting for 73.0% of the total invertebrate benthos (Table 2.6). Ephemeroptera drift occurs mainly (94.5%) at night, with the peak occurring immediately after nightfall (Allan, 1978b).

Certain mayfly behaviors are probably means of dealing with the current. Many Ephemeroptera select for current speeds. Leptophlebiidae avoid currents by inhabiting the peripheral areas of streams. *Paraleptophlebia* spp., detritivore–omnivores, have

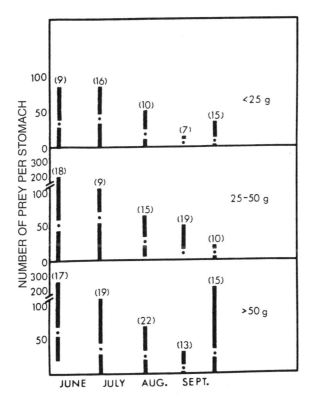

Figure 2.12.   Prey consumption by trout. Mean and range of number of recognizable prey per stomach of trout collected at 3-h intervals over 24 h. Sample size in parentheses. Top panel, trout less than 25 g; middle panel, trout 25 to 50 g; bottom panel, trout greater than 50 g. (From Allan, 1981.)

the ability to burrow into the substrate to avoid torrential changes in the flow. *Baetis* spp. (detritivore–omnivores), *Epeorus* spp. (omnivores), and *Ephemerella* spp. (omnivores) exhibit a progressive movement toward the banks as they grow, thus avoiding the detrimental effects that a stronger midstream current could have on larger bodies (Hynes, 1970).

Ephemeroptera are herbivorous. Most mayflies are either mineral and/or organic scrapers, or deposit-feeding detritivores (Allan, 1975).

Eight species of Ephemeroptera, dominated by *Baetis*, a detritivore–omnivore, and *Rhithrogena*, an omnivore–detritivore, account for 15% of the benthic fauna in Trout Creek (Canton and Ward, 1981).

Ephemeroptera are generally dominant and account for 22 to 56% of the total macroinvertebrate benthos in the Dolores River. Peak populations occur in late summer and early autumn (Fuller and Stewart, 1979).

Baetis bicaudatus (Baetidae), a detritivore–omnivore, is a herbivore scraper, ingesting mineral and organic matter. It ranges in size from 2.5 to 7.0 mm and is known to occur between 3000 and 3100 m altitude (Allan, 1975, 1981, 1982a).

TABLE 2.6. *Composition of Invertebrates in the Stream (Drift and Benthos),*
August 5–6, 1975

	Drift			
	Number/H[a]		Percent Total	Benthos
	Day	Night	Drift	(%)
Ameletus velox	0	124	0.4	0
Baetis bicaudatus	730	12,206	45.7	35.4
Cinygmula sp.	152	2,558	9.6	25.0
Epeorus longimanus	70	1,651	6.1	9.2
Ephemerella coloradensis	2	27	0.1	0.9
E. inermis	2	29	0.1	0.2
Paraleptophlebia vaciva	2	31	0.1	0.4
Rhithrogena spp.	2	65	0.3	1.6
Total Ephemeroptera	961	16,695	62.5	73.0
Alloperla spp.	40	1,028	3.8	5.5
Kogotus modestus	5	17	0.1	0
Pteronarcella badia	14	27	0.1	0
Zapada haysi	120	479	2.1	2.1
Total Plecoptera	182	1,564	6.2	7.7
Brachycentrus sp.	28	39	0.2	0
Rhyacophila spp.	28	93	0.4	1.5
Total Trichoptera	58	140	0.6	1.5
Chironomidae	355	2,198	9.0	1.7
Simulium spp.	923	4,437	18.9	7.7
Total Diptera	1,278	6,635	27.9	9.4
Heterlimnius sp.	41	33	0.3	4.6
Total Coleoptera	53	44	0.4	4.6
Acari	21	5	0.1	0
Oligochaeta	0	0	0	3.7
Total non-Insecta	192	124	1.1	3.8
Emerging aquatic insects	103	103	0.7	0
Total aquatic invertebrates	2,827	25,305	99.4	
Terrestrial invertebrates	99	116	0.6	
Total drift	2,926	25,421		
Number/m² in benthos				1,137

Source: After Allan (1978a).
[a]Numbers drifting per hour are for the entire stream.

B. bicaudatus exhibits certain patterns in stream drift. This species comprises
45.7% of total invertebrate drift and 73% of total Ephemeroptera drift. In all size
classes nocturnal drift exceeds daytime drift. Small nymphs are disproportionately
abundant during the day, while larger nymphs are primarily nocturnal drifters.

B. bicaudatus is among the most abundant insects in rocky riffle habitats. This is
probably due, at least in part, to the streamlining of its body, which makes it ideally

suited for survival in fast currents. Its behavior is probably also an adaptation to current. *B. bicaudatus* nymphs stand high on their legs, with abdomens hanging free in the water, tails swinging side to side in small eddies, and body facing directly into the ever-changing local current. They have the ability to swim very fast in short bursts, with their three tails spread out; the rows of long hairs bordering both sides of the central filament and the inner sides of the outer cerci, make an efficient fin, which moves rapidly up and down. When the insect is at rest, the tails are relaxed and pushed together by the current, with bristles folded. If the current is very swift, the nymphs pull their bodies down so that their ventral sides touch the substratum and they are tucked into the boundary layer (Hynes, 1970).

Baetis tricaudatus (?) (Baetidae), a detritivore–omnivore, occurs in a northern Colorado springbrook (elevation 1597 m) and is very abundant on rubble (Ward and Dufford, 1979).

Baetis spp. (Baetidae), detritivore–omnivores, commonly occur from 2060 to 2606 m altitude in the South Gunnison drainage (Argyle and Edmunds, 1962; Pratt, 1938).

Callibaetis nigritus (Baetidae), an omnivore, is known from a single specimen collected from a northern Colorado springbrook (elevation 1597 m) (Ward and Dufford, 1979).

Pseudocloeon sp. (Baetidae), an omnivore, occurs from 2060 to 2262 m altitude (Argyle and Edmunds, 1962).

Centroptilum sp. (Baetidae), a detritivore–omnivore, is known to occur at 2606 m (Argyle and Edmunds, 1962). Because of its streamlined body, this species is very well adapted to life in fast currents (Hynes, 1970).

Cinygmula sp. (Heptageniidae), an omnivore, ranges in size from 3.0 to 7.5 mm. It is a mineral and organic scraping herbivore (Allan, 1975). This species inhabits fast riffle areas near the shore, from 2060 to 3100 m altitude (Allan, 1981; Argyle and Edmunds, 1962).

Epeorus albertae (Heptageniidae), an omnivore, is found clinging to the surfaces of rubble stones in riffle areas. It occurs from 2060 to 2530 m altitude (Argyle and Edmunds, 1962). Because of its flattened body, this species is particularly well adapted to the current (Hynes, 1970).

Epeorus longimanus (Heptageniidae), an omnivore, ranges from 1.5 to 6.0 mm in size. It is a deposit-feeding detritivore, consuming primarily fine particulate organic matter (FPOM) (Allan, 1975). This species occurs in riffle areas and is ideally suited to the rapid currents in this type of habitat, because of its flattened body. It is found from 2118 to 3100 m altitude. *E. longimanus* is frequently collected with *E. albertae* (Allan, 1981; Argyle and Edmunds, 1962).

Epeorus sp.[6] (Heptageniidae), an omnivore, occurs from 2316 to 2408 m altitude. This species is the strong, clinging type and crawls slowly about on the surfaces of rocks in swift current (Pratt, 1938).

Heptagenia elegantula (Heptageniidae), an omnivore, occurs from 2316 to 2408 m altitude (Pratt, 1938).

---

[6] Pratt (1938) called this genus *Iron*.

Heptagenia solitaria (Heptageniidae), an omnivore, is found between 2060 and 2284 m altitude. It appears to be restricted primarily to the river, with a few ventures into sidestreams. It is most abundant in very rapid riffles (Argyle and Edmunds, 1962).

Heptagenia spp. (Heptageniidae), omnivores, occur between 2060 and 2606 m altitude. These species usually occur in riffle areas but are sometimes found in the slow deep water at the confluences of the Gunnison and its tributaries (Argyle and Edmunds, 1962; Pratt, 1938).

In a northern Colorado springbrook (elevation 1597 m), *Heptagenia* sp., an omnivore, is relatively abundant (Ward and Dufford, 1979).

Rhithrogena elegantula (Heptageniidae), an omnivore–detritivore, ranges in size from 7.0 to 10.0 mm. It is a deposit-feeding detritivore, consuming primarily FPOM (Allan, 1975). With its flattened body, this species is particularly well adapted to fast currents. *R. elegantula* and other members of this genus also have enlarged front gills, which turn forward under the body. This increases the area of marginal contact with the substratum and decreases the amount of water flowing under the nymph, thus giving the insect additional resistance to the current (Hynes, 1970).

Rhithrogena hageni gr. spp. (Heptageniidae), omnivore–detritivores, occur in riffle habitats between 2060 and 3100 m altitude (Allan, 1981; Argyle and Edmunds, 1962).

Rhithrogena robusta (Heptageniidae), an omnivore–detritivore, ranges from 7.5 to 13.5 mm in size. It is a deposit-feeding detritivore, consuming primarily FPOM (Allan, 1975). It is known to occur in torrential habitats, at 2118 and 3100 m altitude (Allan, 1981; Argyle and Edmunds, 1962). Because of its flat body, *R. robusta* is well suited to its environment.

Rhithrogena spp. (Heptageniidae), omnivore–detritivores, occur from 2316 to 2347 m altitude and cling to rocks in swift current (Pratt, 1938).

Ephemerella coloradensis (Ephemerellidae), a herbivore, ranges in size from 3.0 to 7.5 mm. Unlike most Ephemerellidae, it is a herbivore, scraping detritus from rocks (Allan, 1975).

Ephemerella doddsi (Ephemerellidae), a detritivore, occurs from 2118 to 3100 m altitude. This species ranges from 6.5 to 13.5 mm in size. It is a deposit-feeding detritivore, consuming mainly FPOM (Allan, 1981; Argyle and Edmunds, 1962; Pratt, 1938). *E. doddsi* is uncommon, but particularly well adapted to the torrential habitat that is common in the Gunnison drainage. Its flattened thoracic sterna are surrounded by a fringe of outwardly directed hairs that make close contact with the substratum. Its flat ventral surface also has a felt of backwardly directed hairs providing additional contact (Hynes, 1970).

Ephemerella grandis (Ephemerellidae), an omnivore, ranges from 12.5 to 14.5 mm in size. It is a deposit-feeding detritivore, consuming primarily FPOM (Allan, 1975). This species occurs between 2118 and 2548 m altitude. It is commonly found in riffle areas near the bank (Argyle and Edmunds, 1962).

Ephemerella hecuba (Ephemerellidae), an omnivore, occurs from 2198 to 2408 m altitude. It is found under rocks along the shore and is slow moving and well camouflaged in its surroundings. It emerges in late July and August, usually a few at a time, and is probably the last to transform during the summer. Adults are common into September and hover over the water in small swarms during late afternoon. They copulate in the air, and the eggs are discharged by the female directly on the water. This species is restricted to the Gunnison and its larger tributaries (Argyle and Edmunds, 1962; Pratt, 1938).

Ephemerella inermis (Ephemerellidae), an omnivore, ranges from 5.0 to 10.0 mm in size. It is a deposit-feeding detritivore, feeding primarily on FPOM (Allan, 1975). This species appears to be the most widespread mayfly in the drainage. It occurs between 2060 and 2606 m altitude (Argyle and Edmunds, 1962).

E. inermis is also present in a northern Colorado springbrook (elevation 1597 m) (Ward and Dufford, 1979).

Ephemerella infrequens (Ephemerellidae), an omnivore, is known to occur at 3100 m altitude (Allan, 1981).

Ephemerella margarita (Ephemerellidae), an omnivore, occurs between 2197 and 2606 m altitude (Argyle and Edmunds, 1962).

Ephemerella tibialis (Ephemerellidae), an omnivore, is found most commonly in torrential habitats, with large amounts of epilithic algae. It occurs between 2134 and 2286 m altitude (Argyle and Edmunds, 1962).

Ephemerella spp. (Ephemerellidae), omnivores, occur from 2256 to 2408 m altitude (Pratt, 1938).

Paraleptophlebia pallipes (Leptophlebiidae), a detritivore–omnivore, occurs at 2256 m altitude, in a riffle area near the bank (Argyle and Edmunds, 1962). This species has the ability to burrow into the substratum to avoid unfavorable changes in the current (Hynes, 1970).

Paraleptophlebia vaciva (Leptophlebiidae), a detritivore–omnivore, ranges in size from 5.5 to 8.5 mm and is known to occur at 3100 m altitude. It is a herbivorous organic scraper (Allan, 1975, 1981). This species can also avoid torrential currents by burrowing (Hynes, 1970).

Paraleptophlebia sp. (Leptophlebiidae), a detritivore–omnivore, occurs in tributaries between 2118 and 2548 m altitude. This species is most commonly found in riffles, although it occasionally inhabits pools (Argyle and Edmunds, 1962; Pratt, 1938).

Tricorythodes fallax (Tricorythidae), an omnivore, occurs at 2316 m altitude. It appears to prefer the quiet water of pools (Pratt, 1938).

Tricorythodes minutus (Tricorythidae), an omnivore, is found mainly in slow, deep water or in larger tributaries. It occurs primarily in areas where a moderate current moves through vegetated areas, or where there are detritus deposits. This species occurs between 2198 and 2606 m altitude (Argyle and Edmunds, 1962).

Tricorythodes sp. (Tricorythidae), an omnivore, occurs in a northern Colorado springbrook (elevation 1597 m) (Ward and Dufford, 1979).

Ameletus velox (Siphlonuridae), an omnivore–detritivore, ranges from 7.0 to 12.0 mm in size and occurs at 3100 m altitude. It is a suspension feeding herbivore–detritivore (Allan, 1975, 1981).

Ameletus spp. (Siphlonuridae), omnivore–detritivores, occur between 2118 and 2408 m altitude and are well distributed throughout the drainage. They are found in quiet water near riffles, in proximity to the bank (Argyle and Edmunds, 1962; Pratt, 1938).

Siphlonurus occidentalis (Siphlonuridae), an omnivore–carnivore, is found in ponds, in pools that were previously a part of the stream, or in heavily vegetated areas of the stream. Because of the paucity of suitable habitat, this species is not abundant. It occurs between 2201 and 2606 m altitude (Argyle and Edmunds, 1962).

Trichoptera. The caddisflies are among the most numerous insects in the Gunnison River. They have extremely varied morphological and behavioral adaptations, which enable them to function in very harsh stream environments.

Trichoptera have developed a number of ways to survive in rapid currents. Larval and/or pupal cases constructed from sand, pebbles, twigs, and other materials provide ballast and protection in swiftly flowing water. Some caddisflies use silk and other sticky secretions to anchor themselves or their cases in fast currents. Some Psychomyidae construct silk tubes on stone surfaces. Non-case-building Plecoptera, such as *Rhyacophila* spp. (omnivore–carnivores) possess posterior prolegs with clawlike grapples, with which they can cling to the substrata (Hynes, 1970).

Brachycentrus similis (Brachycentridae), an omnivore, occurs between 2316 and 2408 m altitude. This species is common above 2347 m but rare at lower altitudes. It constructs a case from fine fibers of plant material cemented together in such a way as to appear square in cross section. *B. similis* inhabits swift water and attaches its case to an exposed rock surface. Immediately prior to pupation, the larva closes the opening to its case by means of an outwardly swinging trapdoor. Pupae have been observed from June 18 to August 8 (Pratt, 1938).

Brachycentrus sp. (Brachycentridae), an omnivore, ranges in size from 5.0 to 8.5 mm, and is known to occur at 3100 m altitude. It is a suspension-feeding herbivore–detritivore that occasionally switches to carnivory in later instars (Allan, 1975).

Micrasema sp. (Brachycentridae), an omnivore–herbivore, ranges from 4.0 to 8.0 mm in size. It is a suspension-feeding herbivore–detritivore (Allan, 1975).

Glossosoma parvulum (Glossosomatidae), an omnivore, occurs at 2408 m altitude (Pratt, 1938).

Arctopsyche grandis (Hydropsychidae), an omnivore, ranges from 16.0 to 30.0 mm in size. Its dietary habits are essentially omnivorous (Allan, 1975). This species occurs between 2316 and 2408 m altitude. *A. grandis* constructs a rough case from twigs, leaf fragments, and occasionally, small pebbles (Pratt, 1938).

Arctopsyche sp. (Hydropsychidae), an omnivore, occurs between 2347 and 2408 m altitude (Pratt, 1938).

Hydropsyche cockerelli (Hydropsychidae), an omnivore, occurs at 2347 m altitude. It is found under rocks in swift currents and is often associated with *Rhyacophila* spp. (omnivore–carnivore) and *Arctopsyche* sp. (omnivore) larvae. *H. cockerelli* constructs a case from small pebbles and sand (Pratt, 1938).

Hydropsyche morosa (Hydropsychidae), an omnivore, occurs at 2316 m altitude. It is found beneath rocks in rapid currents. This species is frequently found in association with *Rhyacophila* spp. and *Arctopsyche* sp. larvae. It constructs a simple case from small pebbles and sand.

In Piceance Creek *Hydropsyche oslari*, an omnivore, is primarily a large-particle herbivore–detritivore. Animals never comprise more than 25% of the gut content. Chironomids and *Baetis* sp. (omnivore) are dominant food items (Gray and Ward, 1979).

Hydropsyche sp. (Hydropsychidae), an omnivore, is found between 2256 and 2408 m altitude. It seems to prefer swift currents and is usually found beneath rocks (Pratt, 1938).

Parapsyche sp. (Hydropsychidae), an omnivore, ranges in size from 8.0 to 16.0 mm. Its dietary habits are primarily polyphagous (Allan, 1975).

Lepidostoma sp. (Lepidostomatidae), a detritivore, ranges in size from 9.5 to 10.5 mm. It is a large-particle chewer and miner, feeding primarily on detritus (Allan, 1975).

Ecclisomyia conspersa (Limnephilidae), an omnivore, ranges from 10.0 to 13.0 mm in size. It is a large-particle chewer and miner, feeding primarily on detritus (Allan, 1975). Like other Limnephilidae, *E. conspersa* constructs a case from large pebbles and attaches it to rocks in swift water. The exceptionally large size of limnephilid case-building materials is a means of providing necessary ballast in swift currents (Hynes, 1970).

Hesperophylax consimilis (Limnephilidae), a carnivore, is found at 2408 m altitude. It inhabits pools and quiet water along the shore. *H. consimilis* larvae construct cases of small pebbles cemented together in a regular fashion. The cases are usually attached to the substratum at the large, anterior end. Prior to pupation, the caddisfly constructs an addition to the attached end of the case, with larger pebbles. In large populations of *H. consimilis*, the caddisflies occur in close aggregations; this is particularly common during pupation. Pupal mortality in this situation is about 20%. At the conclusion of pupation, adult pupae leave their cases, swim to the surface, crawl onto rocks at the edge of the water, and emerge. Pupation occurs throughout the summer, and pupae have been observed from June 21 through September 13 (Pratt, 1938).

Neothremma alicia (Limnephilidae), a detritivore, ranges in size from 6.0 to 7.0 mm. It is a large-particle chewer and miner, feeding primarily on detritus (Allan, 1975).

Oligophlebodes minutum (Limnephilidae), an omnivore, occurs from 2347 to 2408 m altitude (Pratt, 1938).

Oligophlebodes sp. (Limnephilidae), an omnivore, ranges in size from 5.0 to 8.0 mm. It is a large-particle chewer and miner, feeding primarily on detritus (Allan, 1975).

Psychomyia pulchella (Psychomyiidae), an omnivore, occurs at 2316 m altitude (Pratt, 1938).

Rhyacophila acropedes (Rhyacophilidae), an omnivore–carnivore, ranges in size from 4.0 to 16.0 mm and has carnivorous feeding habits (Allan, 1975). It is known to occur between 3000 and 3200 m altitude. The Rhyacophilidae are free-living during their larval stage; they cling to the substratum in swiftly flowing water and do not build cases. Prior to pupation, however, they construct rough pupal shelters from small stones and silk (Hynes, 1970).

Rhyacophila alberta (Rhyacophilidae), an omnivore–carnivore, ranges in size from 3.5 to 15.5 mm (Allan, 1975).

Rhyacophila angelita (Rhyacophilidae), an omnivore–carnivore, ranges from 3.5 to 10.0 mm in size (Allan, 1975).

Rhyacophila coloradensis (Rhyacophilidae), a carnivore, ranges in size from 7.5 to 13.0 mm (Allan, 1975). It occurs at 2408 m altitude (Pratt, 1938).

Rhyacophila hyalinata (Rhyacophilidae), an omnivore–carnivore, ranges from 7.5 to 13.5 mm in size (Allan, 1975).

Rhyacophila valuma (Rhyacophilidae), an omnivore–carnivore, ranges from 5.0 to 12.5 mm in size and is carnivorous. It occurs between 3000 and 3200 m altitude (Allan, 1975).

Rhyacophila verrula (Rhyacophilidae), an omnivore–carnivore, ranges from 3.5 to 10.5 mm in size (Allan, 1975).

Rhyacophila spp. (Rhyacophilidae), omnivore–carnivores, occur between 2316 and 2408 m and at 3100 m altitude (Allan, 1981; Pratt, 1938).

Twelve species of Trichoptera account for 39% of the insect fauna in Trout Creek. *Brachycentrus americanus* (omnivore) comprises 15% of the trichopteran community. The Lepidostomatidae, dominated by *Lepidostoma moneka* (detritivore), account for about 37% of the caddisfly community. Glossosomatidae, particularly *Glossosoma ventrale* (omnivore) and *Agapetus boulderensis* (omnivore), make up 33% of trichopteran populations. Hydroptilidae, especially *Hydroptila* sp. (herbivore–omnivore), account for 6% of the caddisflies. *Arctopsyche inermis* (omnivore), *Hydropsyche cockerelli* (omnivore), *H. oslari* (omnivore), and other hydropsychids account for only 3% of the trichopteran community (Canton and Ward, 1981).

Hesperophylax incisus (carnivore) occurs on all substrata except sand in a northern Colorado springbrook (elevation 1597 m) and is occasionally found with *Limnephilus frijole* (carnivore). *Hydroptila* sp. (herbivore–omnivore) occurs on all substrata, including sand, and develops large populations on moss. *Hydropsyche* sp. (omnivore) and *Cheumatopsyche* sp. (omnivore) are also present. There is a general downstream increase in the number of trichopteran taxa (Ward and Dufford, 1979).

Diptera. Chironomidae occur over a wide range of altitudes in both quiet and rapid waters. Larvae are often found inhabiting deserted Trichoptera pupal cases, while others construct sand tubes on the undersurfaces of rocks. *Oscillatoria* sp. and *Cladophora* sp. mats often contain large numbers of algae-eating chironomid larvae; the insects frequently appear green in color as a result of the *Cladophora* contained in their digestive tracts. Transformation occurs from late June through late September. Although several species of Chironomidae are undoubtedly represented in the Gunnison River, taxonomic distinctions have not been made (Pratt, 1938).

The Simuliidae are confined to rapid currents. They occur at 2408 m altitude in very large numbers and at 2316 m in small numbers. *Simulium* sp. larvae and pupae are found together, firmly attached to the substratum (Pratt, 1938). They are omnivores.

Simuliidae are well adapted for survival in rapid currents. These insects spin silken mats on the surface of the substratum; they then engage their hooked prolegs in the silk, and thus are well attached, even in the swiftest of currents (Hynes, 1970).

Simuliidae also encounter difficulties when emerging in a rapid current environment. Because the pupa is firmly attached to the substratum, it must emerge under water. It deals with this situation by emerging into a gas bubble, which accumulates just prior to emergence between the pupal and adult cuticles. The bubble then rises rapidly to the surface, and the adult *Simulium* (omnivore) flies off immediately (Hynes, 1970).

The Rhagionidae of the Gunnison River is most probably *Atherix* sp. (carnivore). The larvae are found in swift water over a wide range of altitudes. They occur beneath rocks and are predaceous. The adult females lay their eggs in clusters on the sides of rocks, and then die, still attached to the egg mass. Large egg masses have been observed where hundreds of females oviposited at the same location. After hatching, the larvae fall into the water and are apparently dispersed by the current (Pratt, 1938).

The *Tipula* sp., an omnivore, of Piceance Creek ingest primarily detrital particles greater than 1 mm in longest dimension and thus are classified as *shredders* or large-particle detritivores. Crane flies regularly consume filamentous algae, although this food item comprises less than 20% of the total diet. Although the abundances of filamentous algae differ greatly at two different sites, percent consumption by *Tipula* is similar (Gray and Ward, 1979).

Sixteen species of Diptera account for 20% of the benthic fauna in Trout Creek. *Pericoma* sp., an omnivore, is abundant, and large numbers of chironomids are found associated with growths of *Hydrurus foetidus* (Canton and Ward, 1981).

Dipterans comprise the greatest percentage of total macroinvertebrate biomass on rubble in a northern Colorado springbrook (elevation 1597 m). *Simulium hunteri* (omnivore) is the dominant simuliid; however, *S. latipes* (omnivore) and *S.* nr. *vittatum* (omnivore) are also present. Tipulids comprise the bulk of the dipteran biomass (Ward and Dufford, 1979).

Coleoptera. *Heterlimnius* sp. (Elmidae), an omnivore, ranges in size from 2.5 to 6.5 mm and is presumably a herbivore–detritivore. It accounts for 4.6% of total macroinvertebrate benthos and 0.3% of total stream drift. This species is known to occur at 3100 m altitude (Allan, 1975, 1978b, 1981).

The elmid beetle *Optioservus* sp., an omnivore, occurs in a northern Colorado springbrook (elevation 1597 m), where it is second in abundance to *Baetis*, a detritivore–omnivore. The most diverse coleopteran assemblage occurs on moss (Ward and Dufford, 1979).

The Coleoptera, dominated almost exclusively by *Optioservus seriatus* and *Zaitzevia parvula* (omnivores), comprise 16% of the macroinvertebrate community in Trout Creek (Canton and Ward, 1981).

Fish

The only trout species endemic to the South Gunnison is *Salmo clarki* (Salmonidae), the cutthroat trout, a carnivore. Elimination of its habitat and introduction of non-native competitor fish species have resulted in reductions in the original range of *S. clarki*. Cutthroat trout prefer the cold headwaters of high mountain streams, between 3200 and 3800 m, although they occur at altitudes as low as 1800 m. They inhabit areas with gravel substrata and running water. In small streams the length of cutthroat trout may be no greater than 23 to 26 cm. It is opportunistic in its feeding; aquatic and terrestrial insects and invertebrates are the most common food items; however, small fish, mice, and frogs are also consumed. *S. clarki* spawns in the spring from April to June, depending on water temperatures. Cutthroat trout and rainbow trout hybridize freely if they occur together. Although *S. clarki* lake populations must be maintained by stocking, stream populations are able to reproduce naturally (Allan, 1975, 1978b; Stanford and Ward, 1986).

Salmo trutta (Salmonidae), the brown trout, a carnivore, was introduced into Colorado during the 1890s. It has become well adapted to every major habitat type except cold headwaters above 2990 m, and is usually most common between 2602 and 2900 m. Much of its success is attributable to fall spawning, which usually occurs in October; the young are old enough to withstand high turbidity levels by the time that spring runoff occurs. Spawning usually occurs when the fish reaches 3 or 4 years of age and 20.5 to 38.5 cm in length. Female brown trout usually produce 300 to 1500 eggs at the first spawning. Under favorable conditions, *S. trutta* may live for over 10 years. Brown trout are carnivorous; they consume a wide variety of aquatic and terrestrial organisms, although they prefer insects and large Crustacea. Large trout often seek larger food items, including fish. They are not greatly affected by the late summer decline in macroinvertebrate food items, as many of them are fasting at this time (Allan, 1975, 1978b; Pratt, 1938).

Salmo gairdneri (Salmonidae), the rainbow trout, a carnivore, was introduced to Colorado around 1882. It prefers water temperatures of 10 to 15°C, and is known to occur at 2670 m altitude. Rainbow trout spawn during the spring in silt-free gravel riffles. They are omnivorous, typically eating insects, crustaceans, and sometimes algae; large trout occasionally feed on smaller fish (Allan, 1975).

The principal food items for the rainbow trout during the first three weeks of June are *Pteronarcys* sp. nymphs (omnivore–carnivores) in the lower Gunnison River and stoneflies and mayflies in the upper river. During late June and July, adult and larval Ephemeroptera and Trichoptera are dominant in the diet, throughout the drainage. From late July through September there is a significant decline in the abundance of rainbow trout prey; at this time *S. gairdneri* consumes large quantities of *Cladophora* sp. (Pratt, 1938).

Salvelinus fontinalis (Salmonidae), the brook trout, a carnivore, was introduced into Colorado in the late nineteenth century. This species is tolerant of a certain amount of temperature fluctuation (from just above 0 to 25°C); however, it apparently prefers water temperatures between 12 and 18°C. Although found at lower altitudes, it flourishes in the Upper Montane and Subalpine Regions between 2743 and 3283 m altitude, particularly above 2900 m (Allan, 1975, 1978b, 1981, 1982b).

Brook trout spawn in the fall, usually in October. One-year-old males are generally sexually mature, whereas females do not produce ova until at least age 2. Female brook trout produce between 300 and 1000 eggs, depending on their size. Usually about 90% of the eggs will survive to hatching; however, as few as 5% of the fry may survive their first summer. Natural mortality of adult fish varies widely and is affected significantly by environmental conditions and harvesting.

Brook trout feed primarily on aquatic and terrestrial insects; Crustacea may also constitute a portion of the diet (Nelson, 1971). The temporal pattern of feeding activity of *S. fontinalis* varies seasonally. During June and July, feeding activity is greatest in early evening. By August, periodicity is less pronounced; however, late afternoon feeding is apparent. In early September feeding is aperiodic, and in late September, feeding is greatest at midday (Allan, 1981).

There is also a significant seasonal decline in number of prey consumed for each trout size class (Figure 2.12). Biomass ingested also decreases seasonally in smaller trout; however, there appears to be no change in consumed biomass for large brook trout. This seasonal decline in number of prey ingested is consistent with seasonal decline in prey densities (Allan, 1981).

Many food items occur in similar proportional abundances in brook trout diets and stream drift (Figure 2.13). Ephemeroptera (*Baetis* sp. (detritivore–omnivore), *Cinygmula* sp. (omnivore), *Rhithrogena* sp. (omnivore–detritivore), and Ephemerellidae, and Diptera (primarily Simuliidae) are dominant (Figure 2.14). Simuliids and *Baetis* sp. are commonly ingested during June and July. Surface drift insects (terrestrial and emerging) are consumed most heavily in August and September (Allan, 1981).

Other food items tend to be either over- or underrepresented in the brook trout diet, compared to stream abundances. The general pattern appears to be that trout select for larger prey items; however, general abundance in the stream is still the primary influence on abundance in brook trout diets (Allan, 1981, 1982b).

Catostomus latipinnis (Catostomidae), the flannelmouth sucker, a carnivore, is restricted in Colorado to the Colorado River drainage, below 2500 m altitude. Adults inhabit quiet pools; however, they forage in swifter waters, especially at night. Young fish are found more frequently in shallower water. Flannelmouth suckers apparently

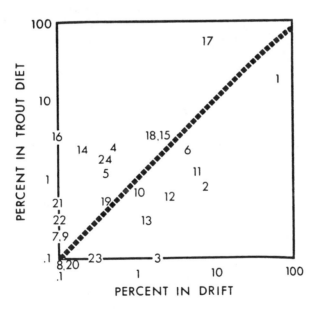

Figure 2.13.   Relationship between percent composition of prey items consumed by trout and percent composition of drift. Data of June 6–7, 1977. Total recognizable prey from the stomachs of 44 trout = 2135. Spearman's r_s = 0.46, $p < 0.05$. 1, *Baetis bicaudatus*; 2, *Cinygmula* sp.; 3, *Epeorus longimanus*; 4, *Rhithrogena hageni* and *R. robusta*; 5, *Ephemerella infrequens*; 6, *E. coloradensis*; 7, *E. doddsi*; 8, *Paraleptophlebia vaciva*; 9, *Ameletus velox*; 10, *Alloperla* spp.; 11, *Zapada haysi*; 12, Perlodidae; 13, other Plecoptera; 14, *Brachycentrus* sp.; 15, *Rhyacophila* spp.; 16, other Trichoptera; 17, Simuliidae; 18, Chironomidae; 19, other Diptera; 20, *Heterlimnius* sp.; 21, other Coleoptera; 22, Acari; 23, emerging aquatic insects; 24, terrestrial invertebrates; 25, trout eggs. (From Allan, 1981.)

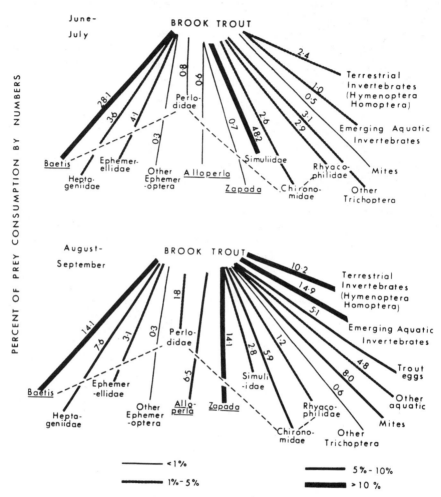

FOOD WEB   CEMENT CREEK

Figure 2.14.   Food web in Cement Creek in early and late summer, based on stomach analysis of trout and stoneflies (Plecoptera) over seasons. (From Allan, 1982b.)

spawn in the summer. They are herbivorous and feed primarily on algae, diatoms, and vascular plants (Pratt, 1938; Stanford and Ward, 1986).

Catostomus discobolus (Catostomidae), the bluehead sucker, a carnivore, occurs most commonly between 2000 and 3400 m; however, it does occur in lower reaches.

Pantosteus delphinus (Catostomidae), the mountain sucker, a carnivore, thrives in small, cold, headwater streams, although it is also found in larger streams and rivers. This species has a pronounced notch between the upper and lower lips and a hard cartilaginous ridge on the edge of the jaw. Unlike other suckers, these morphological characteristics allow the mountain sucker to scrape algae, diatoms, and organic material off rocks. This food resource niche partitioning allows *P. delphinus* to coexist with other sucker species. Mountain suckers spawn between May and August in riffle areas of the stream or at the lower ends of deep holes. After hatching, the fry occupy shallow areas in large schools (Pratt, 1938).

Rhinichthys osculus[7] (Cyprinidae), the speckled dace, a carnivore, occurs in the South Gunnison River at 1500 m and between 2256 and 3200 m; it is less common in the upper reaches. It occupies riffle areas and spawns in late spring. *Cladophora* sp. and *Potamogeton* provide shelter for the fry. *R. osculus* seldom reaches a length greater than 13 cm (Pratt, 1938; Stanford and Ward, 1986).

Cottus bairdi (Cottidae), the mottled sculpin, a carnivore, occurs in small mountain trout streams at altitudes between 2200 and 3700 m. It inhabits the bottoms of cold trout streams, usually hiding under stones. The sculpin spawns in the spring, attaching the fertilized eggs in clumps of 100 to 500 eggs under rocks. The male guards the eggs. This species feeds on insects, small fish, and algae (Stanford and Ward, 1986).

Xyrauchen texanus (Catostomidae), the razorback sucker, a carnivore, is endemic to the Upper Colorado River Basin. However, its range has been reduced in recent years because of alterations in flows and impeded access to spawning streams by dams. The razorback sucker is known to occur at the mouth of the Yampa River and in a flooded gravel pit near Grand Junction, Colorado. Although *X. texanus* is frequently found in larger streams, the gravel/rock riffles of headwater streams still provide essential habitat for a portion of the razorback sucker life cycle. Diet consists primarily of algae, plankton, plant debris, and Ephemeroptera, Trichoptera, Diptera, and Chironomidae larvae (McAda and Wydoski, 1980).

Xyrauchen texanus is known to occur in Colorado in the Yampa, Colorado, San Juan, and Gunnison Rivers (Figure 2.15) (Behnke and Benson, 1980).

Three hybrids have also been collected from this area: *Xyrauchen texanus* × *Catostomus latipinnis* (carnivore), *Catostomus latipinnis* × *Catostomus commersoni* (carnivore), and *Catostomus commersoni* × *Catostomus discobolus* (carnivore) (McAda and Wydoski, 1980).

Ptychocheilus lucius (Cyprinidae), the Colorado squawfish (carnivore); *Gila cypha* (Cyprinidae), the humpback chub (omnivore); and *Gila elegans* (Cyprinidae), the bonytail chub (omnivore) are also present in the upper Colorado drainage (McAda and Wydoski, 1980).

Ptychocheilus lucius, a carnivore, is known to occur in Colorado in the Yampa, White, Gunnison, and Colorado Rivers (Figure 2.16) (Behnke and Benson, 1980).

Gila cypha, an omnivore, is known to occur in Colorado in the Yampa and Colorado Rivers (Figure 2.17) (Behnke and Benson, 1980).

The four fish species native to the Dolores River system—*Gila robusta* (omnivore), *Rhinichthys osculus* (carnivore), *Catostomus latipinnis* (carnivore), and *Catostomus discobolus* (carnivore)—are commonly found in medium-sized tributaries of the upper Colorado River. The chub, dace, and flannelmouth sucker dominate the

---

[7]Pratt (1938) called this species *Apocope yarrowi.*

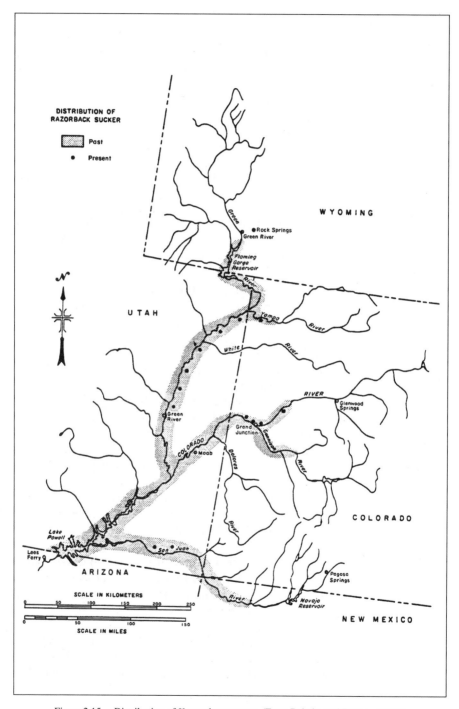

Figure 2.15.   Distribution of *Xyrauchen texanus*. (From Behnke and Benson, 1980.)

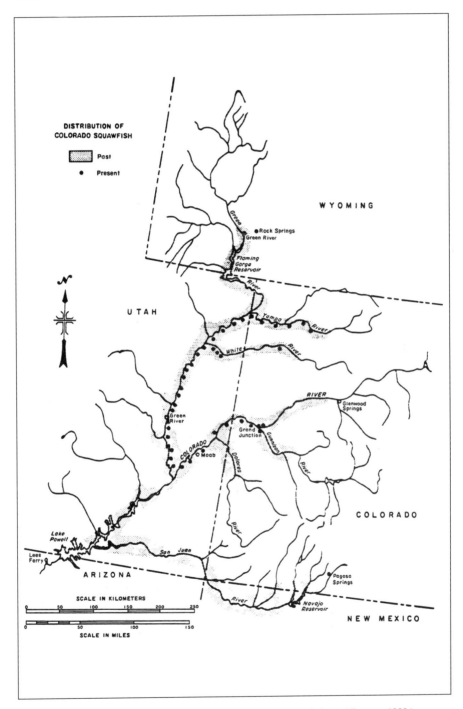

Figure 2.16. Distribution of *Ptychocheilus lucius*. (From Behnke and Benson, 1980.)

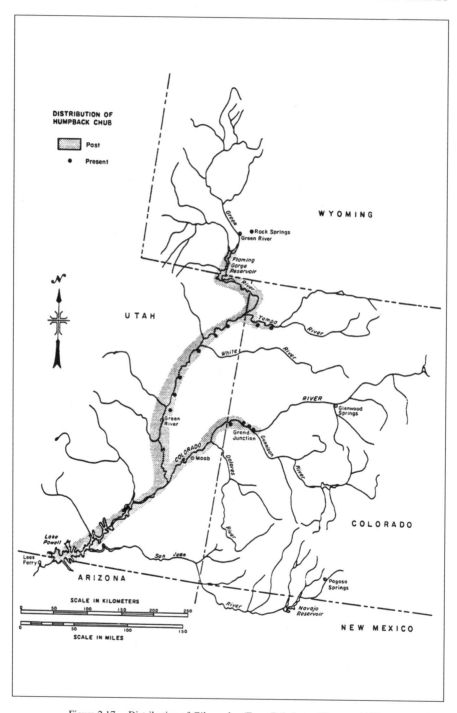

Figure 2.17.   Distribution of *Gila cypha*. (From Behnke and Benson, 1980.)

fish fauna in the dry, middle section of the river. Bluehead suckers are found only in the upper and lower sections, since they favor larger riffles (Holden and Stalnaker, 1975b).

ASSOCIATIONS OF SPECIES FOUND AT VARIOUS ALTITUDES IN COLORADO HARD HEADWATERS

The South Gunnison River drainage ranges from 1800 to 4200 m altitude and transverses four altitudinal biotic zones:

1. *Alpine Tundra Region:* >3400 m
2. *Subalpine Region:* 2850 to 3350 m
3. *Upper Montane Region:* 2450 to 2750 m
4. *Lower Montane Region:* 2060 to 2350 m

The interregional altitudinal reaches are transitional zones, where some overlap of characteristic biota may occur.

The biological communities of the Gunnison can be classified into distinct groups on the basis of habitat and intraspecies associations. The hierarchy of these groups is as follows (Pratt, 1938):

1. *Association:* a climax community of a definite taxonomic composition, with uniform physiognomy and habitat conditions
2. *Society:* a subgroup of an association
3. *Associes:* a developmental community of the same rank as an association
4. *Socies:* a subgroup of an associes

The Alpine Tundra Region

The Alpine Tundra Region of the Gunnison River drainage occurs above 3400 m altitude in several Gunnison headwater tributaries (Tomichi Creek, Slate River, East River, Cement Creek, and Taylor River). In this region the primary habitat is white water. This habitat is typically characterized by a definitive type of community (Pratt, 1938).

The white-water habitat is characterized by the *Simulium* socies, omnivores. No other animals are found regularly associated with *Simulium* in the Gunnison River. *Acroneuria* spp. (carnivores) sometimes wander into the *Simulium* socies, but are transient residents. *Salmo gairdneri* and *Salmo trutta*, carnivores, occasionally forage in white-water areas (Pratt, 1938).

Algae.   The plant community of the Alpine Tundra Region consists entirely of diatoms and some scattered blue-green algae. Insect abundances decrease drastically in the Alpine Tundra Region (Figure 2.14). Although stoneflies, mayflies, caddisflies, and blackflies are present, the diversity and abundances are quite low.

Salmo clarki, the cutthroat trout (carnivore), prefers the cool headwaters and occurs at altitudes up to 3800 m. *Cottus bairdi*, the mottled sculpin (carnivore), also prefers cool mountain streams and is present at altitudes up to 3700 m. *Catostomus discobolus*, the bluehead sucker (carnivore), occurs near the lower limit of the Alpine Tundra Region.

The Subalpine Region

In the South Gunnison River the Subalpine Region ranges from about 2850 to 3350 m altitude. White water and clear rapids are the two basic types of permanent habitat. The white-water areas contain the characteristic *Simulium* socies, which are omnivores.

The clear rapids habitat has a rocky substratum and is characterized by the *Hydropsyche–Acroneuria–Ephemerella–Salmo* associes at high altitudes, which are omnivore–carnivore associations. Organisms that are able to deal successfully with rapid currents comprise this community (Pratt, 1938).

The most abundant invertebrates inhabiting areas between and beneath stones are *Hydropsyche* sp. (omnivore), some rhyacophilids, *Ephemerella* sp. (omnivore), *Baetis* sp. (detritivore–omnivore), Chironomidae, and Rhagionidae. Subdominants are *Pteronarcella badia* (omnivore–carnivore), *Heptagenia elegantula* (omnivore), and *Arctopsyche* sp. (omnivore) (Pratt, 1938).

The dominant animals clinging to the exposed tops and sides of stones are *Brachycentrus similis* (omnivore), certain rhyacophilids, *Acroneuria pacifica* (carnivore), *Claassenia* sp. (carnivore), *Epeorus* sp. (omnivore), and *Rhithrogena* sp. (omnivore–detritivore). Chironomidae are found seasonally among *Oscillatoria* filaments; *Baetis* sp. (omnivore–detritivore) and *Ephemerella* sp. (omnivore) nymphs sometimes inhabit *Scleropodium obtusifolium* mats (Pratt, 1938).

The trout dominate the free-swimming stratum of this associes. In the Subalpine Region, *Salvelinus fontinalis* and *Salmo gairdneri*, which are carnivores, appear to be the most common fish species (Pratt, 1938).

The plant community of the Subalpine Region is dominated by a small number of algal species. The blue-green algae *Nostoc* sp. and *Oscillatoria* sp. occur occasionally. Diatoms are the most diverse class of algae and are dominated by *Synedra ulna*, *Diatoma hiemale*, and *Navicula* sp. The moss *Scleropodium obtusifolium* occurs in this region; however, *Cladophora* and aquatic vascular plants do not.

Helobdella stagnalis, a carnivore, is one of the only noninsect invertebrates that occupies the Subalpine Region. It occurs up to 3200 m altitude, on a variety of substrata. Possibly because of its small size, this leech is able to avoid current shear. It is also able to tolerate temperatures as low as 5°C (Herrmann, 1970).

The Plecoptera are fairly common in the Subalpine Region. *Pteronarcella badia* (omnivore–carnivore) occurs in relatively quiet areas in leaf masses and debris accumulations. *Arcynopteryx signata* (carnivore), *Isogenus elongatus* (carnivore), *I. modestus* (carnivore), *Diura knowltoni* (carnivore), *Isoperla patricia* (carnivore), *Kogotus modestus* (carnivore), *Megarcys signata* (carnivore), *Alloperla borealis* (omnivore–carnivore), *A. coloradensis* (omnivore–carnivore), *Alloperla pallidula* (omnivore–carnivore), *Paraperla frontalis* (omnivore), *Acroneuria pacifica* (carnivore), *Claassenia sabulosa* (carnivore), *Nemoura oregonensis* (detritivore), *N. besametsa* (detritivore), *Acroneuria* sp. (carnivore), and *Zapada haysi* (detritivore) occur beneath rocks and rubble in riffle areas.

The Ephemeroptera of the Subalpine Region are abundant. *Ephemerella infrequens* (omnivore), *E. doddsi* (omnivore), *Baetis bicaudatus* (detritivore–omnivore), *Epeorus longimanus* (omnivore), *Rhithrogena hageni* gr. spp. (omnivore–detritivores),

R. robusta (omnivore–detritivore), *Paraleptophlebia vaciva* (detritivore–omnivore), *Ameletus velox* (omnivore–detritivore), and *Heptagenia elegantula* (omnivore) occupy riffle areas between and beneath rocks.

The trichopteran community in this region is composed of *Hydropsyche* sp. (omnivore), *Rhyacophila acropedes* (omnivore–carnivore), *R. valuma* (omnivore–carnivore), *Arctopsyche* sp. (omnivore), and *Brachycentrus similis* (omnivore).

Chironomidae and Rhagionidae are the most common dipterans at high altitudes. *Heterlimnius* sp. (omnivore) is apparently the only coleopteran in this zone.

Salmo clarki (carnivore), *S. trutta* (carnivore), *Salvelinus fontinalis* (carnivore), *Pantosteus delphinus* (carnivore), *Catostomus discobolus* (carnivore), *Rhinichthys osculus* (carnivore), and *Cottus bairdi* (carnivore) comprise the fish community of the Subalpine Region of the Gunnison River.

The Upper Montane Region

The Upper Montane Region of the South Gunnison River ranges from about 2450 to 2750 m. In this region, as in the Subalpine Region, the *Hydropsyche–Acorneuria–Ephemerella–Salmo* associes (omnivore–carnivore–omnivore–carnivore) is prevalent in the clear rapids areas (Pratt, 1938).

The pool areas are characterized by the general *Catostomus–Pantosteus–Salmo* association (carnivores). These areas are generally the most permanent habitats found and therefore usually contain climax communities. The *Catostomus–Pantosteus–Salmo* association can be subdivided into societies on the basis of pool depth; in this region generally only the shallow pool society is found (Pratt, 1938).

Pools less than 0.6 m in depth are typically of the *Hesperophylax*–small *Catostomus*–small *Pantosteus*–small *Salmo* (carnivore) society. The substrata are usually composed of stones and, to a lesser degree, silt. *Hesperophylax consimilis*, a herbivore–detritivore, and Chironomidae may occur attached to rocks or unattached on a sandy bottom. Small *Catostomus latipinnis* (carnivore) and *Pantosteus delphinus* (carnivore) are generally more common than trout (Pratt, 1938).

Scleropodium obtusifolium is common on rocks in the Upper Montane Region.

Glossiphonia complanata, a carnivore, and *Erpobdella punctata*, a carnivore, are found in the lower reaches of the Upper Montane Region, to an altitude of about 2500 m. *Helobdella stagnalis*, a carnivore, occurs throughout this zone on a variety of substrata.

The Plecoptera communities of this region are very diverse. The majority of stoneflies are found among rocks in areas of rapid current. *Isogenus aestivalis* (carnivore), *I. elongatus* (carnivore), *I. modestus* (carnivore), *Paraperla frontalis* (carnivore), *Claassenia sabulosa* (carnivore), *Arcynopteryx signata* (carnivore), *A. parallela* (carnivore), *Isoperla fulva* (carnivore), *I. patricia* (carnivore), *I. pinta* (carnivore), *Diura knowltoni* (carnivore), and *Acroneuria pacifica* (carnivore) are all found in these riffle areas. *Pteronarcella badia* (omnivore–carnivore) is found in quiet habitats, usually among leaf and debris accumulations. *Alloperla signata* (omnivore–carnivore) occurs in deep holes and appears to be rare. *Eucapnopsis brevicauda* (detritivore), *Capnia coloradensis* (detritivore), *C. gracilaria* (detritivore), *C. limita* (detritivore), *C. logana* (detritivore), *C. poda* (detritivore), *Nemoura besametsa* (detritivore), *N. cinctipes* (detritivore), *N. coloradensis* (detritivore), *Brachyptera pacifica* (detritivore), *Alloperla borealis* (omnivore–carnivore), *A. coloradensis* (omnivore–carnivore), *A. pallidula* (omnivore–carnivore), *Leuctra sara* (detritivore), and *Kogotus modestus* (carnivore) are also present in the Upper Montane Region.

Most of the Ephemeroptera occur in rocky riffle habitats in fast currents. *Baetis* sp. (detritivore–omnivore), *Centroptilum* sp. (detritivore–omnivore), *Epeorus albertae* (omnivore), *Heptagenia* spp. (omnivores), *Ephemerella grandis* (omnivore), *Ephemerella inermis* (omnivore), *Ephemerella margarita* (omnivore), and *Paraleptophlebia* sp. (detritivore–omnivore) occur at or below 2606 m altitude. *Cinygmula* sp. (omnivore), *Epeorus longimanus* (omnivore), *Rhithrogena hageni* gr. spp. (omnivore–detritivores), *R. robusta* (omnivore–detritivore), and *Ephemerella doddsi* (omnivore) occur throughout the Upper Montane Region.

Tricorythodes minutus (omnivore) and *Siphlonurus occidentalis* (omnivore–carnivore) occur in deep pools or among vegetation in slow currents. Both species are found up to an altitude of 2606 m.

The fish of the Upper Montane Region are diverse. *Salmo clarki* (carnivore) occurs occasionally throughout this region in gravel riffles, although it prefers higher altitudes. *S. trutta* (carnivore) occurs above 2602 m, associated with a variety of habitats. *S. gairdneri* (carnivore) is known to occur at 2670 m, in silt-free gravel riffles. *Salvelinus fontinalis* (carnivore) occurs in the upper altitudinal reaches of this region. *Catostomus latipinnis* (carnivore) occurs near the lower altitudinal limits, below 2500 m; it occupies quiet pools, although it frequently forages in swifter waters. *C. discobolus* (carnivore), *Rhinichthys osculus* (carnivore), and *Cottus bairdi* (carnivore) occur throughout the Upper Montane Region. *Catostomus discobolus* (carnivore) is usually found in quiet waters. Adult *R. osculus* occupy riffle areas, while the fry are often associated with vegetation. *C. bairdi* is generally discovered under stones in running water.

The Lower Montane Region

The Lower Montane Region of the South Gunnison River ranges from 1800 to approximately 2350 m. At this lower altitude the clear rapids habitat is usually characterized by the *Pteronarcys–Rhagionidae–Rhinichthys–Salmo* (omnivore–carnivore/carnivore/carnivore/carnivore) associes. In this region, *Helodrilus* sp., a herbivore–detritivore, is found in the hyporheic zone, where it has burrowed into the gravel. *Hydropsyche* sp. (omnivore), *Pteronarcys californica* (omnivore–carnivore), *Ephemerella* sp. (omnivore), *Baetis* sp., Chironomidae, Rhagionidae, and *Rhinichthys osculus* (carnivore) are found among stones. *Acroneuria pacifica* (carnivore), *Claassenia depressa* (carnivore), *Epeorus* sp. (omnivore), *Rhithrogena* sp. (omnivore–detritivore), and *Rhyacophila* sp. (omnivore–carnivore) are the dominant species on rocks. Chironomidae occur seasonally in *Cladophora* mats. Trout are found free-swimming in the current (Pratt, 1938).

In permanent pools the *Hesperophylax*–small *Catostomus*–small *Pantosteus*–small *Salmo* society is common at depths of less than 0.6 m. In the Lower Montane Region, *Potamogeton* sp. is abundant and often serves as a refuge for the numerous small suckers (Pratt, 1938).

In pools of moderate depths (0.6 to 1.8 m) the large *Catostomus*–large *Pantosteus*–Chironomidae society is found. Large suckers are extremely abundant in great schools. Trout are rare, except when driven from shallower water by drought (Pratt, 1938).

In pools greater than 1.8 m deep, the Chironomidae society is found. Chironomids are virtually the only invertebrates present. Suckers and trout are rare (Pratt, 1938).

The shallow-water habitat along the edges of clear rapids in the Lower Montane Region possesses a rather definitive biota. The small *Salmo*–small *Catostomus*–small *Pantosteus* socies is found in this temporary habitat. These areas are populated by species moving in from permanent habitats. The fish are from the clear-water rapids and seem to prefer these areas for foraging. *Hesperophylax consimilis* (herbivore–detritivore), *Physa* sp. (omnivore), and *Glossiphonia complanata* (omnivore) emigrate to this area from pools. *Ephemerella hecuba* (omnivore) is apparently a permanent resident of these marginal areas and burrows into the mud during the winter (Pratt, 1938).

The blue-green algae *Nostoc* sp. and *Oscillatoria* sp. occur in this region primarily on rocks. The green alga *Cladophora* sp. is abundant on rocks in swiftly flowing water. *Ulothrix* sp., *Mougeotia* sp., *Oedogonium* sp., *Spirogyra* sp., *Stigonema* sp., *Microspora crassior*, *Prasiola* sp., and the desmids *Euastrum* sp., *Cosmarium undulatum*, and *Closterium* sp. are also present.

The chrysophytes *Hydrurus foetidus* and *Vaucheria* sp. are present in this zone. Diatom communities are dominated by *Diatoma* sp. and *Synedra ulna*.

Potamogeton sp. is quite common in the Lower Montane Region, forming large beds on sand or silt substrata.

Mollusca are common in this region. *Ancylus* sp., a herbivore–detritivore, and *Physa* sp., an omnivore, occur in shallow rapids near the shore and in pooled areas. *Pisidium casertanum*, a detritivore, occurs below 2300 m in mud and/or associated with *Potamogeton*. *Pisidium compressum* is less common and occurs only at about 2300 m; it inhabits mud bottoms in shallow water.

The annelid *Helodrilus* sp., a herbivore–detritivore, is common in the Lower Montane Region of the South Gunnison. It is generally found at least 10 in. beneath a gravel substratum.

The leeches *Erpobdella punctata* (carnivore), *Hellobdella stagnalis* (carnivore), and *Glossiphonia complanata* (carnivore) are present in this region. *E. punctata* and *H. stagnalis* are found in a wide range of habitats. *G. complanata* occurs most commonly under rocks.

The plecopteran communities of the Lower Montane Region are quite diverse. *Isocapnia crinita* (detritivore), *Brachyptera pacifica* (detritivore), *B. pallida* (detritivore), *Capnia gracilaria* (detritivore), *C. limita* (detritivore), *C. logana* (detritivore), *C. poda* (detritivore), *Nemoura cinctipes* (detritivore), *Isoperla fulva* (carnivore), *I. modesta* gr. sp. (carnivore), *I. mormona* (carnivore), *I. patricia* (carnivore), *I. pinta* (carnivore), *I. quinquepunctata* (carnivore), *Leuctra sara* (detritivore), *Paraperla frontalis* (omnivore), *Alloperla coloradensis* (omnivore–carnivore), *A. pallidula* (omnivore–carnivore), *A. signata* (omnivore–carnivore), *Isogenus aestivalis* (carnivore), *I. elongatus* (carnivore), *Claassenia depressa* (carnivore), *Claassenia sabulosa* (carnivore), *Pteronarcys californica* (omnivore–carnivore), *Arcynopteryx parallela* (carnivore), *A. signata* (carnivore), *Diura knowltoni* (carnivore), *Acroneuria pacifica* (carnivore), *Skwala parallela* (carnivore), and *Pteronarcella badia* (omnivore–carnivore) occur in this region.

The ephemeropteran community of the Lower Montane Region consists of *Baetis* sp. (detritivore–omnivore), *Pseudocloeon* sp. (omnivore), *Cinygmula* sp. (omnivore), *Epeorus albertae* (omnivore), *E. longimanus* (omnivore), *Heptagenia elegantula* (omnivore), *H. solitaria* (omnivore), *Heptagenia* sp. (omnivore), *Rhithrogena hageni* gr. sp. (omnivore–detritivore), *R. robusta* (omnivore–detritivore), *Ephemerella doddsi* (omnivore), *E. grandis* (omnivore), *E. hecuba* (omnivore), *E. inermis* (omnivore), *E. margarita* (omnivore), *E. tibialis* (omnivore), *Paraleptophlebia pallipes* (detritivore–omnivore), *Tricorythodes fallax* (omnivore), *T. minutus* (omnivore), *Ameletus* sp. (omnivore–detritivore), and *Siphlonurus occidentalis* (omnivore–carnivore).

The Trichoptera known to occur in this region are *Brachycentrus similis* (omnivore), *Glossosoma parvulum* (omnivore), *Arctopsyche grandis* (omnivore), *Arctopsyche* sp. (omnivore), *Hydropsyche cockerelli* (omnivore), *H. morosa* (omnivore), *Hydropsyche* sp. (omnivore), *Hesperophylax consimilis* (herbivore–detritivore), *Oligophlebodes minutum* (omnivore), *Psychomyia pulchella* (omnivore), and *Rhyacophila* spp. (omnivore–carnivore).

Chironomids occur in both quiet and rapid water habitats in the Lower Montane Region and are frequently found inhabiting *Cladophora* mats. Simuliids and rhagionids occur firmly attached to rocks in rapid currents.

The fish communities of the region are composed of *Rhinichthys osculus* (carnivore), *Cottus bairdi* (carnivore), a few wide-ranging *Salmo clarki* (carnivore), and the suckers *Catostomus latipinnis* (carnivore) and *C. discobolus* (carnivore).

ASSOCIATIONS OF SPECIES WITH VARIOUS SUBSTRATA

Habitats of the various species are identified in Table 2.3. Examples of organisms in various habitats are given below.

Vegetation and Debris

For example, found associated with vegetation were the blue-green alga *Entophysalis incrustans* and the euglenophytes *Euglena* sp. and *Phacus* sp. The diatom *Synedra*, and the chlorophytes *Characium* sp., *Closterium* sp., *Cosmarium undulatum*, *Cosmarium* sp., *Euastrum* sp., and *Micrasterias* sp. were also found in vegetated habitats.

The protozoan *Paramecium* sp. (omnivore), the planarian *Dugesia dorotocephala* (detritivore), and the erpobdellid *Dina dubia* (carnivore) were frequently found associated with vegetation and debris. The cladocerans *Simocephalus* sp. (omnivore–detritivore) and *Chydorus* sp. (omnivore–detritivore) were found in similar habitats.

Two elmids *Heterlimnius* sp. (omnivore) and *Optioservus* sp. (omnivore) occurred in areas that were vegetated. Two dipterans, the psychoidid *Pericoma* sp. (omnivore) and the tipulid *Limonia* sp. (herbivore), were also found associated with vegetation and debris. The ephemerellid mayfly *Ephemerella tibialis* (omnivore) occupied vegetated areas, as did the pteronarcid stonefly *Pteronarcys californica* (omnivore–carnivore) and the limnephilid caddisfly *Hesperophylax incisus* (herbivore–detritivore).

The gastropod *Physa* sp. (omnivore) was frequently found in habitats with aquatic macrophytes.

Several fish were found in vegetated habitats: the salmoniid *Prosopium williamsoni* (carnivore), the cyprinids *Cyprinus carpio* (omnivore) and *Pimephales promelas* (detritivore–herbivore), and the centrarchid *Lepomis cyanellus* (carnivore).

Sand

For example, the macrophytes *Potamogeton* sp. and *Ranunculus* sp. frequently grew in sandy habitats. Diatoms are also common. The planarian *Dugesia dorotocephala* (detritivore), the lumbricid *Eiseniella tetraedra* (herbivore–detritivore), the naidids *Paranais* sp. (herbivore–detritivore) and *Pristina* sp. (herbivore–detritivore), and the hirudid *Haemopsis marmorata* (carnivore) were also frequently associated with sandy substrates.

The chironomid *Prodiamesa* sp. (omnivore), the heptageniid mayflies *Heptagenia* spp. (omnivores), and the hydroptilid caddisflies *Hydroptila* spp. (herbivore–omnivores) were the few insects documented from sandy areas. Several mollusc taxa were found in sandy substrata: the physid *Physa* sp. (omnivore), the lymnaeids *Lymnaea* spp. (omnivores), and the sphaeriid *Pisidium* sp. (detritivore).

Mud and Silt

For example, the euglenophytes *Euglena* sp. and *Phacus* sp., the diatoms *Synedra* spp., and the desmids *Closterium cucumis*, *Closterium* sp., *Cosmarium undulatum*, *Cosmarium* sp., *Euastrum* sp. and *Micrasterias* sp. were found associated with mud and silt habitats. The chaosid *Amoeba* sp., an omnivore, was also found in muddy and silty areas.

Potamogeton sp. and *Ranunculus* sp. grew in mud substrates. The planarian *Dugesia dorotocephala* (detritivore), the lumbricid *Eiseniella tetraedra* (herbivore–detritivore), the naidids *Paranais* sp. (herbivore–detritivore) and *Pristina* sp. (herbivore–detritivore), the glossiphoniid *Hellobdella stagnalis* (carnivore), and the erpobdellid *Erpobdella punctata* (carnivore) were also found associated with mud and silt habitats.

The leptophlebiid mayflies *Paraleptophlebia pallipes* (detritivore–omnivore) and *P. vaciva* (detritivore–omnivore) occurred in muddy areas. Two stonefly taxa, the chloroperlid *Alloperla signata* (omnivore–carnivore) and the pteronarcid *Pteronarcella badia* (omnivore–carnivore) were also found associated with mud and silt substrates. Several caddisflies, the hydroptilids *Hydroptila* spp. (herbivore–omnivores) and the limnephilids *Hesperophylax incisus* (herbivore–detritivore) and *Limnephilus frijole* (omnivore) occupied mud and silt habitats.

Several molluscs were found in muddy areas: the physid *Physa* sp. (omnivore), the lymnaeid *Lymnaea* spp. (omnivores), the planorbid *Gyraulus* sp. (herbivore–detritivore), and the sphaerids *Pisidium casertanum* (detritivore), *Pisidium compressum* (detritivore), and *Pisidium* sp. (detritivore).

Several fish were also documented from mud and silt habitats: the cyprinid *Cyprinus carpio* (omnivore) and the ictalurids *Ictalurus melas* (carnivore) and *I. punctatus* (carnivore).

Gravel

For example, the planarian *Dugesia dorotocephala*, a detritivore, was found associated with gravel substrata. Several annelids were also found in this type of habitat: the lumbricids *Eiseniella tetraedra* (herbivore–detritivore) and *Helodrilus* sp. (herbivore–detritivore), the glossiphoniid *Hellobdella stagnalis* (carnivore), and the erpobdellid *Erpobdella punctata* (carnivore). Several caddisflies were also found associated with gravel bottoms: the hydroptilids *Hydroptila* spp. (herbivore–omnivore), and the limnephilids *Hesperophylax incisus* (herbivore–detritivore) and *Limnephilus frijole* (omnivore).

The molluscs *Physa* sp. (omnivore), *Lymnaea* spp. (omnivores), *Gyraulus* sp. (herbivore–detritivore), and *Pisidium* sp. (detritivore) were documented from gravel substrates. A number of fish species were found in this type of habitat also: the salmonids *Salmo clarki* (carnivore), *S. gairdneri* (carnivore), *S. trutta* (carnivore), and *Salvelinus fontinalis* (carnivore), the catostomid *Xyrauchen texanus* (carnivore), the cyprinids *Gila cypha* (omnivore), *G. elegans* (omnivore), *G. robusta* (omnivore), *Notropis lutrensis* (carnivore–omnivore), *N. stramineus* (carnivore–omnivore), and *Rhinichthys osculus* (carnivore).

Cobble and Rubble

For example, nearly all algae with documented habitats were found in cobble and rubble habitats. All of the Nostocaceae, most of the Oscillatoriaceae, *Rivularia* sp., and *Stigonema* sp. comprised most of the blue-green algae populations found on these substrates. The euglenophytes *Euglena* sp. and *Phacus* sp. were also associated with cobble and rubble. A number of chrysophytes were found in cobble and rubble habitats, including the Xanthophyceae *Vaucheria germinata* and *Vaucheria* sp., the Chrysophyceae *Hydrurus foetidus*, and the diatoms *Melosira* sp., *Diatoma hiemale*, *Diatoma* sp., *Fragilaria capucine*, *Fragilaria* spp., *Meridion* sp., *Synedra* spp., *Tabellaria* sp., *Achnanthes lanceolata*, *A. minutissima*, *Cocconeis pediculus*, *C. placentula*, *Cymbella* spp., *Gomphonema ehrengregii*, *Gomphonema* spp., *Navicula radiosa*, *N. rhynchocephala*, *N. viridula*, *Navicula* spp., *Epithemia gibba*, and *Nitzschia amphibia*.

A number of green algae were also found associated with cobble and rubble substrata: *Characium* sp., *Ulothrix zonata*, *Ulothrix* spp., *Microspora* sp., *Chaetophora* sp., *Draparnaldia plumosa*, *Stigeoclonium* sp., *Closterium* sp., *Mougeotia* sp., *Spirogyra* sp., *Zygnema* sp., *Cladophora glomerata*, *Cladophora* sp., *Enteromorpha* sp., *Oedogonium* sp., and *Prasiola* sp. The red alga *Batrachospermum monoliforme* was also found growing on cobble and rubble.

The peritrichid *Vorticella* sp. was documented from cobble and rubble habitats. The bryophyte *Scleropodium obtusifolium* was found growing on cobble and rubble also.

Several annelids were found associated with cobble and rubble: the lumbricid *Eiseniella tetraedra* (herbivore–detritivore), the glossosiphoniids *Glossiphonia complanata* (carnivore), *Hellobdella stagnalis* (carnivore), and *Placobdella ornata* (carnivore), and the erpobdellid *Erpobdella punctata* (carnivore). The amphipod *Gammarus* sp. (omnivore) was also found in cobble and rubble habitats.

Several coleopterans were found on cobble and rubble: *Deronectes* sp. (carnivore), *Heterlimnius* sp. (omnivore), *Optioservus seriatus* (omnivore), *Optioservus* sp. (omnivore), and *Zaitzevia parvula* (omnivore). A very large number of dipterans, trichopterans, ephemeropterans, and plecopterans were also found on these substrates (see Table 2.3).

Several molluscs were found in cobble and rubble habitats as well: the gastropods *Physa* sp. (omnivore), *Lymnaea* spp. (omnivores), and *Ancylus* sp. (herbivore–detritivore), and the pelecypod *Pisidium* sp. (detritivore).

A number of fish were documented from cobble and rubble stretches of stream: the salmonids *Salmo clarki* (carnivore), *S. gairdneri* (carnivore), *S. trutta* (carnivore), and *Salvelinus fontinalis* (carnivore); the esocid *Esox lucius* (carnivore); the catostomids *Catostomus commersoni* (carnivore), *C. discobolus* (carnivore), *C. latipinnis* (carnivore), and *Xyrauchen texanus*; the cyprinids *Cyprinus carpio* (omnivore), *Gila cypha* (omnivore), *G. elegans* (omnivore), *G. robusta* (omnivore), *Pimephales promelas* (detritivore–herbivore), *Ptychocheilus lucius* (carnivore), *Rhinichthys osculus* (carnivore), and *Richardsonius balteatus* (omnivore); the centrarchids *Lepomis cyanellus* (carnivore), *L. macrochirus* (carnivore), and *Micropterus salmoides* (carnivore); the percid *Stizostedion vitreum* (carnivore); the cottid *Cottus bairdi* (carnivore); and the ictalurids *Ictalurus melas* (carnivore) and *I. punctatus* (carnivore).

STRUCTURE OF FUNCTIONAL ECOSYSTEMS

Alpine Tundra Region: 3400 Meters

The plant community of the Alpine Tundra Region consists entirely of diatoms and some scattered blue-green algae. The structure of the ecosystem is given in Table 2.7.

The omnivores were blackflies, particularly *Simulium socies,* and mayflies.

TABLE 2.7. *Structure of Functioning Ecosystem: Alpine Tundra Region, Hardwater Ecosystem (Tomichi Creek, Slate River, East River, Cement Creek, and Taylor River)*

————————————Carnivores————————————
Acroneuria spp.
Catostomus discobolus
Salmo clarki
S. gairdneri
S. trutta

————————————Omnivores————————————
Ephemeroptera
Simulium socies

—————————Detritus—————————	—————————Algae—————————
	Blue-green algae
	Diatoms

The carnivores were several species of *Acroneuria* and caddisflies, which are mostly omnivorous, and the fish *Salmo gairdneri, S. trutta,* and *S. clarki* and *Catostomus discobolus*. Diversity and abundance are quite low, although species of stoneflies, mayflies, caddisflies, and blackflies are present as noted above.

Diatoms and blue-greens were obviously eaten by the omnivores, which in turn were consumed by the carnivores.

Subalpine Zone: 2850 to 3350 Meters

The Subalpine Zone varies in altitude from 2850 to 3350 m and is composed of white water and clear rapids. These are the two permanent habitats. The structure of the ecosystem is given in Table 2.8.

The plant community consists of blue-green algae, *Nostoc* spp. and *Oscillatoria* sp. The diatoms are the most diverse, being dominated by *Synedra ulna, Diatoma hiemale,* and *Navicula*.

The moss, *Scleropidium obtusifolium,* is fairly common in this region, occurring on boulders and rocks.

The detritivores were *Nemoura oregonensis, N. besametsa,* and *Zapada haysi*.

Omnivores that were present were *Paraperla frontinalis, Ephemerella infrequens, E. doddsi, Epeorus longimanus, Heptagenia elegantula,* and the caddisflies *Hydropsyche* sp., *Arctopsyche,* and *Brachycentrus similis*. Of the chironomids that were omnivores, *Heterlimnius* sp. was apparently the only coleopteran or beetle in this area.

Omnivore–carnivores that were present consisted of *Pteronarcella badia,* which was found particularly in quiet areas in leaf masses and debris. Also present were *Alloperla borealis, A. coloradensis,* and *A. pallidula* and the trichopterans *Rhyacophila acropedes* and *R. valuma*.

The carnivores that dominated the open areas were the trout *Salvelinus fontinalis* and *Salmo gairdneri,* and *Hellobdella stagnalis*. The stonefly *Arcynopteryx signata* was also a carnivore, as were *Isogenus elongatus, I. modestus, Diura knowltoni, Isoperla patricia, Kogotus modestus, Megarcys signata, Acroneuria pacifica,* and *Claassenia sabulosa*. Carnivores were numerous and indicate that the species in the lower stages of the food web must have been common to support such a large number of carnivorous species. Other carnivores present were *Salmo clarki, S. trutta, Salvelinus fontinalis, Pantosteus delphinus, Catostomus discobolus, Rhinichthys osculus,* and *Cottus bairdi*. Thus the fish community was diverse and well established in this zone, indicating that their food sources, which were probably omnivores and omnivore–carnivores and detritivores, must have been fairly well established.

Lower Montane Region: 1800 to Approximately 2350 Meters

In the Lower Montane Region there is a greater variety of habitats than one finds at higher altitudes. This is probably because the substrates are more variable. In the higher altitudes, boulders, rocks, rubble, and gravel are typically the substrates. In this area, the Lower Montane Region, there are some boulders, cobbles, and rocks but there are also areas of mud and sand where the current is slow, as in pools, and there

TABLE 2.8. *Structure of Functioning Ecosystem: Subalpine Zone, Hardwater Ecosystem (South Gunnison River, 2850–3350 m Altitude)*

	Detritivores	Detritivore–herbivores	Detritivore–omnivores	Herbivores	Herbivore–omnivores
Carnivores	*Acroneuria pacifica* *Arcynopteryx signata* *Catostomus discobolus* *Claassenia sabulosa*	*Cottus bairdi* *Diura knowltoni* *Hellobdella stagnalis* *Isogenus elongatus*	*I. modestus* *Isoperla patricia* *Kogotus modestus* *Megarcys signata*	*Pantosteus delphinus* *Rhinichthys osculus* *Salmo clarki*	*S. gairdneri* *S. trutta* *Salvelinus fontinalis*
Omnivore–carnivores		*Alloperla borealis* *A. coloradensis*	*A. pallidula* *Pteronarcella badia*	*Rhyacophila acropedes* *R. valuma*	
Omnivores		*Arctopsyche* sp. *Brachycentrus similis* *Epeorus longimanus*	*Ephemerella doddsi* *E. infrequens* *Heptagenia elegantula*	*Heterlimnius* sp. *Hydropsyche* sp. *Paraperla frontinalis*	
Detritivores	*Nemoura besametsa* *N. oregonensis* *Zapada haysi*				
Herbivores				*Oscillatoria* sp. *Synedra ulna*	
Algae				*Diatoma hiemale* *Navicula* spp. *Nostoc* spp	
Detritus					

are also areas where there is relatively clean sand and gravel. The structure of the ecosystem is given in Table 2.9.

The areas where the current is swift and clear are usually characterized by a detritivore–herbivore, *Hellodrillus* sp., and an omnivore–detritivore, an unidentified species of the mayfly, *Baetis* sp. Also present, particularly associated with the rocks, was an omnivore–detritivore, *Rhithrogena* sp. The omnivores in this area, particularly in the gravel, were unidentified specimens of *Hydropsyche* sp., *Ephemerella* sp., and *Epeorus* sp. The omnivore–carnivores in this area were *Pteronarcys* sp., *Pteronarcys californica*, *Rhyacophila* sp., and *Epeorus* sp. These were more common among the rocks.

The carnivores present in this association were *Rhinichthys osculus*, which was found among the stones. Other insects common on the rocks that were carnivores were *Acroneuria pacifica* and *Claassenia depressa*. Other carnivores that were present in these low mountain areas, particularly associated with permanent pools, were the catfish *Catostomus* sp., *Pantosteus* sp., and *Salmo* sp. In this region the *Potamogeton* sp. was abundant and formed a habitat for many insects and a hiding place for fish to feed. In the pools of moderate depth the large *Catostomus* sp., a carnivore, was found, as was *Pantosteus*, a carnivore. These are chironomids and were in the bed material of the pools.

In shallow-water habitats along the edges of clear rapids, several fish were found. They are all carnivores. They were small specimens of *Salmo* sp., of *Catostomus* sp., and of *Pantosteus* sp. They were found mainly in temporary habitats. These areas are populated from species moving in from permanent habitats. Associated with the clear rapids–preferring areas for foraging were a herbivore–detritivore, *Hesperophylax consimilis*, and the omnivorous *Physa* sp. Also into this area, having migrated from pools, were the carnivorous *Glossosiphonia complanata* and the mayfly *Ephemerella hecuba*, which is an omnivore and appears to be a permanent resident of these marginal areas.

Rocks, Rubble, and Gravel.   Blue-green algae were common on rocks, particularly species belonging to the genera *Nostoc* sp. and *Oscillatoria* sp. Green algae that were present on rocks in swiftly moving water were unidentified species of *Cladophora*. In less rapid water were several other green algae: the filamentous *Ulothrix* sp., *Mougeotia* sp., *Oedogonium* sp., *Spirogyra* sp., *Stigonema* sp., *Microspora* sp. *Crassior* sp., and *Prasiola* sp. Also present in this habitat associated with swiftly moving water near rocks (they must also have been in areas where the current was reduced) were the desmids, *Euastrum* sp., *Cosmarium undulatum*, and *Closterium* sp. These algae undoubtedly formed the base of the food web for many of the invertebrates and fish and therefore would be the primary producers.

Other algae forming in this group are the chrysophytes and the diatoms. The chrysophytes were *Hydrus foetidus* and *Vaucheria* sp. The diatom communities were dominated by *Diatoma* sp. and *Synedra ulna*. Quite common, forming sometimes large beds on sand and silt substrate, was *Potamogeton* sp.

Molluscs were fairly common in this area. They were found in shallow rapids near the shore and in pooled areas. They were snails belonging to the genera *Ancylus* sp. and *Physa* sp. The former is a herbivore–detritivore and the *Physa* is believed to be an omnivore.

TABLE 2.9. *Structure of Functioning Ecosystem: The Lower Montane Region, Hardwater Ecosystem (South Gunnison River, 1800 to Approximately 2350 m Altitude)*

Carnivores

Acroneuria pacifica	*Catostomus* sp.	*Erpobdella punctata*	*Pantosteus* sp.	*Salmo clarki*
Catostomus. discobolus	*Claassenia depressa*	*Glossosiphonia complanata*	*Rhinichthys osculus*	*Salmo* sp.
Catostomus latipinnis	*Cottus bairdi*	*Helobdella stagnalis*		

Omnivore—carnivores

Pteronarcys californica
Pteronarcys sp.
Rhyacophila sp.

Omnivores

Arctopsyche grandis	*Epeorus* sp.	*E. tibialis*	*Hydropsyche cockerelli*	*Physa* sp.
Arctopsyche sp.	*Ephemerella doddsi*	*Ephemerella* sp.	*H. morosa*	*Pseudocloeon* sp.
Brachycentrus similis	*E. grandis*	*Glossosoma parvulum*	*Hydropsyche* sp.	*Psychomyia pulchella*
Cinygmula sp.	*E. hecuba*	*Heptagenia elegantula*	*Oligophlebodes minutum*	*Tricorythodes fallax*
Epeorus albertae	*E. inermis*	*H. solitaria*	*Paraperla fontinalis*	*T. minutus*
E. longimanus	*E. margarita*	*Heptagenia* sp.		

TABLE 2.9. *Continued*

Detritivores			Herbivores	
Detritivores	**Detritivore–herbivores**	**Detritivore–omnivores**	**Herbivores**	**Herbivore–omnivores**
Brachyptera pacifica	*Ancylus* sp.	*Baetis* sp.		
B. pallida	*Helodrilus* sp.	*Rhithrogena* sp.		
Capnia gracilaria	*Hesperophylax consimilis*			
C. limita				
C. logana				
C. poda				
Isocapnia crinita				
Leuctra sara				
Nemoura cinctipes				
Pisidium casertanum				
P. compressum				

Detritus		Algae	
		Cladophora spp.	*Oedogonium* sp.
		Closterium sp.	*Oscillatoria* sp.
		Cosmarium undulatum	*Potamogeton* sp.
		Crassior sp.	*Prasiola* sp.
		Diatoma sp.	*Spirogyra* sp.
		Euastrum sp.	*Stigonema* sp.
		Hydrus foetidus	*Synedra ulna*
		Microspora sp.	*Ulothrix* sp.
		Mougeotia sp.	*Vaucheria* sp.
		Nostoc spp.	

(continues)

Occurring in mud or associated with the aquatic plant *Potamogeton* was *Pisidium casertanum*, a detritivore, and *P. compressum*, a detritivore. This was somewhat less common than *P. casertanum*.

In this Lower Montane Region in the South Gunnison was found a worm, *Helodrilus* sp., a herbivore–detritivore. Quite a few leeches, of wide-ranging habitats usually in association with rocks, were in this area. They were *Erpobdella punctata*, a carnivore; *Helobdella stagnalis*, a carnivore; and *Glossosiphonia complanata*, a carnivore.

The plecopteran community consists of species belonging in various stages of the food web. The detritivores in the Lower Montane Region are mainly stoneflies such as *Isocapnia crinita*, *Brachyptera pacifica*, *B. pallida*, *Capnia gracilaria*, *C. limita*, *C. logana*, *C. poda*, *Nemoura cinctipes*, and *Leuctra sara*. Also found was the herbivore–detritivore, *Hesperophylax consimilis*. Omnivorous species occurring in this general habitat were much more common. They consisted of *Paraperla fontinalis*, a stonefly, and the mayflies: *Pseudocloeon* sp., *Cinygmula* sp., *Epeorus albertae*, *E. longimanus*, *Heptagenia elegantula*, *H. solitaria*, *Heptagenia* sp., *Ephemerella doddsi*, *E. grandis*, *E. hecuba*, *E. inermis*, *E. margarita*, *E. tibialis*, *Tricorythodes fallax*, and *T. minutus*. Other omnivores were the trichopterans *Brachycentrus similis*, *Glossosoma parvulum*, *Arctopsyche grandis*, and *Arctopsyche* sp., a caddisfly, which is also an omnivore, as are many other caddisflies, such as *Hydropsyche cockerelli*, *H. morosa*, *Oligophlebodes minutum*, and *Psychomyia pulchella*.

Chironomids were found in both quiet and rapid water at this altitude. They were also common in *Cladophora* mats. They sometimes are found firmly attached to rocks in rapid currents. The carnivores of this community were mainly fish, although some carnivorous insects were present. Also present were the leeches *Erpobdella punctata*, *Helobdella stagnalis*, and *Glossosiphonia complanata*. The fish were *Rhinichthys osculus*, *Cottus bairdi*, *Salmo clarki*, and the suckers *Catostomus latipinnis* and *C. discobolus*.

SUMMARY

The hardwater streams differ from the softwater streams in that the schists, gneisses, and granites are overlain with sandstone, limestone, and cretaceous shale. The Gunnison River drainage exemplifies hard headwater streams of the Colorado Rocky Mountain area. The Gunnison is bounded by West Elk Mountains on the north, Elk Mountains on the northeast, the Sawatch Mountains on the east, Cochetops Hills on the southeast, and Uncompahgre Mountains on the south and west.

The flow of the South Gunnison is typically rapid and turbulent. In the area where the river has a rather steep drop (i.e., 2.84 m/km^{-1}) the water flow is swift and the bottoms are usually made of well-rounded cobbles of various sizes with occasional rocks or boulders. The bottom of the Gunnison River is composed of large rubble and boulders. Sand and silt become more common in a downstream direction. At the junctions of the Gunnison and many of its tributaries, the water slows down and sand and gravel become the bottom substrate. The annual mean temperature increases as one progresses downstream. The daily thermal gain is about 6°C from the headwaters to the Colorado River.

The south fork of the Gunnison and its tributaries fall into what Pennak terms hard to very hard streams. Their total alkalinity usually ranges from 100 to 220 mg L^{-1}. Dissolved oxygen in the Gunnison River ranges from 7.4 to 9.5, with dissolved oxygen saturation ranging from 90 to 115%. pH generally ranges between 7.8 and 8.6. Calcium is the dominant ion and often is linked with sulfates. Nitrogen as N is relatively low in the Gunnison River, varying from 0 to 1 mg L^{-1} or ppm. Ammonia is detectable throughout most of the drainage, although the concentration is very low. The system appears to be nutrient limited in the headwaters, which often happens in cool mountain streams. Suspended solids range from 3 to 11 mg L^{-1}.

Parachute Creek is another very hardwater stream of the upper Colorado River, and the river drainage basin is typically hardwater throughout. The White River basin is also a hardwater stream being underlain by oil shale, halite, nahcolite, and gypsum. The North Fork of the White is also hard, with an average CO_2 value of 59.7 ppm. Other hardwater streams are the Yampa River and the Dolores River basin.

Primary production in these hard headwater streams is caused by diatoms, filamentous blue-green algae, unicellular green algae, and small bits of leaves, conifer needles, and detritus. Phytoplankton is usually not very well developed. The total organic carbon pool in the Gunnison increased from 1 to 10 mg L^{-1} on an average from headwaters to mouth.

Piceance Creek contains a fairly large amount of allochthonous material, particularly at the higher altitudes. Detritus seems to be the main source of organic matter in the food web. *Cladophora* sp. has a low energy content, 1.63 ± 0.68 kcal/g, whereas allochthonous organic matter has an energy content of 3.28 ± 0.46 kcal g^{-1}. The algae are the most important primary producers. They are species of *Nostoc*, *Oscillatoria*, *Rivularia*, *Stigeonema*, *Ulothrix*, *Cladophora*, and diatoms. Despite a large amount of autotrophy, heterotrophic conditions are dominant, indicating that algae respiration and the action of bacteria and fungi on organic detritus of the substrate more than balance the algae photosynthesis.

The riparian vegetation along the Colorado hard headwaters is mainly sagebrush, willows, cottonwood, grasses of various types, alder, birch, hawthorn, wild rose, some conifers, poison ivy, serviceberry, dogwood, currants, and raspberries. The algae on rocks is dominantly a species of *Nostoc*. Midge larvae are often found associated with the *Nostoc*. It also occurs in marginal pools. *Oscillatoria* sp. often forms slimy masses on rock surfaces, particularly in late summer. Common green algae in the South Gunnison River drainage are *Cladophora*, *Stigeoclonium*, and *Ulothrix*, with *Cladophora* being the most commonly observed green alga. The diatoms are the most diverse group of algae in the Gunnison and are very common. The moss *Scleropodium obtusifolium* is abundant in the upper South Gunnison drainage. *Potamogeton* sp. is one of the most characteristic plants in the lower Gunnison. Protozoans have occurred in colonies under rocks, are often associated with dead leaves in marginal pools, and are sometimes found to be common in colonies of *Spirogyra* sp.

Several species of mollusc, such as *Pisidium caesertanum*, *P. compressum* are often found associated with *Potamogeton* and other algae. Molluscs such as *Physa* and *Ancyclus* are common on stones in the shallow shore rapids in the lower river. A number of annelids that are herbivore–detritivores occur in gravel between rocks and swift rapids, particularly an unidentified species of *Helodrillus*. Also present in plant beds is a herbivore–detritivore, *Eiseniella tetraedra*. Crustacea such as *Gammarus* sp. and *Hyallela azteca*, which are common, occur in certain substrates such as pool areas occasionally.

The stoneflies of a certain group seem to be limited to low altitudes, others to high altitudes, and others occur over a wide range of altitudes. The effect of altitudes seems to be mainly one of temperature and, indirectly, food, which often is not as abundant at high altitude. Stoneflies of the suborder *Filipalpia* are usually herbivorous, while nymphs of the suborder *Setipalpia* are usually carnivores. The relative increase in carnivores at high altitudes is probably due to the reduced amount of algae in these elevations. Current is also an important factor. Medium-sized streams often provide the greatest variety of stoneflies, probably because they have the greatest variety of habitats and resources. Omnivore–carnivores seem to be fairly common in this stream and alter its percentage of animal-to-plant food, depending on what is available. Another important factor for stoneflies is their habitat, some of them preferring swift waters, and some of them preferring slower waters between stones and relatively slow water on the downstream side of a stone.

As contrasted to the stoneflies, which are often more carnivorous in their food preferences, many species of the Ephemeroptera are omnivores or omnivore–detritivores. Mayflies seem to prefer habitats of varying amounts of current. Thus even on a given rock, where the current may be very variable, different species will be found where the current is turbulent, where it is laminar, or where there is practically no current at all.

Caddisflies are among the most numerous insects in the Gunnison River. They have varied morphological and behavioral adaptations which allow them to function in a variety of stream habitats. The caddisflies are found in a variety of habitats where the current is variable. They are most common where the current is moderately swift.

The Diptera, particularly the chironomids, are found in a great variety of habitats. Many of them have algae as their main food source. The Simulidae were confined to rapid currents and are often found on the surface of rocks and boulders. The tipulids digest mainly detritus, although some of them consume a fair amount of algae. They are typically found where the current is moderate. For example, dipterans comprise the greatest percentage of the total macroinvertebrate biomass on rubble in springbrooks that have moderately high elevations. However, they are common in many different habitats.

Only one species of trout was endemic to the South Gunnison. It was *Salmo clarki*. One trout, *Salmo trutta*, is a carnivore and is fairly common in these mountainous streams. They feed largely on macroinvertebrates, and if they decline, the population of fish also declines. Most of the trout seem to be carnivorous. However, they may at times be omnivorous.

Various species of algae, invertebrates, and fish seem to be more common at different altitudes and thus form distinctive associations. The Alpine Tundra Region is usually above 3400 m, the Subalpine Region between 2850 and 3350 m, the Upper Montane Region between 2450 and 2750 m, and the Lower Montane Region from 2060 to 2350 m.

The structure of functional ecosystems at various altitudes varies both in available habitat and in types of organisms. In the Alpine Tundra Region, the substrates are largely bolders and the flow is that of plunge pools because of the high gradient. The numbers of species forming the ecosystem are relatively few. At the base of the ecosystem are diatoms and some blue-green algae. Omnivores feeding on the algae and perhaps other insects are blackfly larvae, particularly *Simulium* socies and mayflies, which are mostly omnivores. A few carnivores are present. All in all the ecosystem consists of relatively few species, those that typically are found in these kinds of habitats.

In the Subalpine Region, the gradient is less steep; not only are rocks and rubble present but also sand. Growing on the surface of the rocks is the moss *Scleropodium obtusifolium*. Intermixed with the moss are the blue-green algae *Nostoc* sp. and *Oscillatoria* sp. Growing on the surface of the rocks but also associated with the moss are several species of diatoms. The detritivores consisted of three species, two belonging to the genus *Nemoura* and one to the genus *Zapada*. Mayflies were fairly common, but a few species of caddisflies were also present. They tended to be omnivorous in food habits. Associated with the sand and gravel in this habitat were chironomids, which are probably omnivores. In leaf masses and debris was an omnivore–carnivore, *Pteronarcella badia*. Also present were several species of *Alloperla*, the stonefly genus, and two species of trichopterans. The carnivores were dominated in the open areas by trout, particularly *Salvelinus fontinalis* and *Salmo gairdneri*. Also present was *Helobdella stagnalis*.

Stoneflies were also present. As seen by the charts of the structure of the ecosystem (Tables 2.7, 2.8, and 2.9), it is very evident that they are much more diverse than at the higher altitudes. The algae were mainly diatoms and blue-green algae. Several detritivores were present, with omnivore–carnivores and carnivores being the most common species in the ecosystem.

In the Lower Montane Region there was a greater variety of habitats than in the higher altitudes. This is probably because the substrate is more variable. Whereas in the higher altitudes, boulders, rocks, and rubble or gravel are typical substrate, in this area where the current is slower, the gradient is less, there still were some rocks and cobbles and a few boulders, but there were also areas of mud and sand where the current formed quiet pools. Where the current was swift and clear, *Helodrillus* sp., a detritivore–herbivore, and some unidentified species of mayflies that were probably omnivore–detritivores were present, such as those that belong to the genus *Baetis* sp. As seen from the charts, there was a much greater diversity of algae forming the base of the food web. Also, the detritivores were much more common, probably feeding on detritus of moss and leaves from the watershed as well as perhaps remnants of algal filaments. As one would expect, the omnivore–carnivores would be much more present, due to the higher numbers of organisms in the base of the food web (i.e., the algae and the detritivores). Carnivores were also quite diverse, consisting mainly of fish. Several fish were found along the edges of the stream in rapidly flowing water. Most of the fish were carnivores. Associated with the clear rapids, preferring such areas to forage, were herbivore–detritivores such as *Hesperophylax consimilis* and an omnivorous snail, *Physa* sp. In the rocks, rubble, and gravel were blue-green algae, also several species of green algae and diatoms. The diatom community was dominated by *Diatoma* sp. and *Synedra ulna*. Molluscs were fairly common in this area. They were mainly in shallow rapids and in pool areas and were species belonging to the genera *Ancylus* and *Physa*.

As stated above, not only were the habitats more diverse but also the organisms that occupied these habitats. There was a well-developed fauna in this reach, consisting of herbivore–detritivores, detritivores, a great many species of omnivores, some species that were omnivore–carnivores, and a well-developed carnivorous fauna. This no doubt was caused by the greater diversity of habitats. The richer faunas, which then produce more food for various organisms in the food web as well as causing more algae, such as *Potamogeton*, to be present, produced shelter and also had definite attachment habitats for many species.

Since substrates of mud and sand were present, worms were more common. As one would expect, the plecopteran community was well developed, and these species were present in various stages in the food web, being detritivores, herbivore–detrivores, omnivores, and carnivores.

BIBLIOGRAPHY

Allan, J. D. 1975. The distributional ecology and diversity of benthic insects in Cement Creek, Colorado. Ecology 56: 1040–1053.

Allan, J. D. 1978a. Trout predation and the size composition of stream drift. Limnol. Oceanogr., 23(6): 1231–1237.

Allan, J. D. 1978b. Diet of brook trout *Salvelinus fontinalis* Mitchill and brown trout *Salmo trutta* L. in an alpine stream. Int. Ver. Limnol. Verh., 20(3–4): 2045–2050.

Allan, J. D. 1981. Determinants of diet of brook trout *Salvelinus fontinalis* in a mountain stream. Can. J. Fish. Aquat. Sci., 38(2): 184–192.

Allan, J. D. 1982a. Feeding habits and prey consumption of three setipalpian stoneflies (Plecoptera) in a mountain stream. Ecology 63(1): 26–34.

Allan, J. D. 1982b. The effects of reduction in trout density on the invertebrate community of a mountain stream. Ecology 63(5): 1444–1455.

Argyle, D. W., and G. F. Edmunds. 1962. Mayflies (Ephemeroptera) of the Curecanti Reservoir basins. Univ. Utah Anthropol. Pap., 59(8): 178–189.

Behnke, R. J., and D. E. Benson. 1980. Endangered and threatened fishes of the upper Colorado River basin. Colorado State Univ. Bulletin 503A.

Canton, S. P., and J. V. Ward. 1981. The aquatic insects with emphasis on Trichoptera of a Colorado stream affected by coal strip mine drainage. Southwest. Nat. 25(4): 453–460.

Ellis, M. M. 1914. Fishes of Colorado. Univ. Colo. Stud. 11(1): 1–136.

Fuller, R. L., and K. W. Stewart. 1977. The food habits of stoneflies (Plecoptera) in the upper Gunnison River, Colorado. Environ. Entomol. 6(2): 293–302.

Fuller, R. L., and K. W. Stewart. 1979. Stonefly (Plecoptera) food habits and prey performance in the Dolores River, Colorado. Am. Midl. Nat. 101(1): 170–181.

Gray, L. J., and J. V. Ward. 1979. Food habits of stream benthos at sites of differing food availability. Am. Midl. Nat. 102(1): 157–167.

Herrmann, S. J. 1970. Systematics, distribution, and ecology of Colorado Hirudinea. Am. Midl. Nat. 83:1–37.

Hill, W. R. 1929. Microscopic zoology of Boulder Creek. Master's thesis. Univ. Colorado.

Holden, P. B., and C. B. Stalnaker. 1975a. Distribution and abundance of mainstream fishes of the middle and upper Colorado River basins, USA. Trans. Am. Fish. Soc. 104(2): 217–231.

Holden, P. B., and C. B. Stalnaker. 1975b. Distribution of fishes in the Dolores and Yampa River systems of the Upper Colorado basin. Southwest. Nat. 19(4): 403–412.

Hynes, H. B. N. 1970. The ecology of running waters. Liverpool University Press. Liverpool.

Johnson, M. D. 1972. The aquatic biota of St. Vrain Creek and the South Platte River adjacent to the Fort St. Vrain nuclear generating station. Master's thesis. Colorado State Univ.

Jordan, D. S. 1889. Report of Explorations in Colorado and Utah during the summer of 1889, with an account of the fishes found in each of the river basins examined. Bull. U.S. Fish Comm. 9: 1–40.

Knight, A. W. 1965. Studies on the stoneflies of the Gunnison River drainage in Colorado. Doctoral dissertation. Univ. Utah.

Knight, A. W., and D. W. Argyle. 1962. Limited limnological studies of the Gunnison River, Colorado. Univ. Utah Anthropol. Pap. 59(8).

Knight, A. W., and A. R. Gaufin. 1966. Altitudinal distribution of stoneflies (Plecoptera) in a Rocky Mountain drainage system. J. Kans. Entomol. Soc. 39: 668–675.

Knight, A. W., and A. R. Gaufin. 1967. Stream type selection and association of stoneflies in a Colorado river drainage system. J. Kans. Entomol. Soc. 40: 347–352.

Li, H. W. 1968. Fishes of the South Platte River basin. Master's thesis, Colorado State Univ.

McAda, C. W., and R. S. Wydoski. 1980. The razorback sucker *Xyrauchen texanus* in the upper Colorado River basin. U.S. Fish and Wildlife Service Technical Paper 0(99): 1–16.

Pennak, R. W. 1943. Limnological variables in a Colorado mountain stream. Am. Midl. Nat. 29:186–199.

Pennak, R. W. 1971a. A fresh-water archiannelid from the Colorado Rocky Mountains. Trans. Am. Microsc. Soc. 90(3): 372–375.

Pennak, R. W. 1971b. Toward a classification of lotic habitats. Hydrobiologia 38(2): 321–334.

Pennak, R. W. 1977. Trophic variables in Rocky Mountain trout streams. Arch. Hydrobiol. 80: 253–285.

Pennak, R., and J. W. Lavelle. 1979. In situ measurements of net primary productivity in a Colorado Mountain stream. Hydrobiologia 66(3): 227–235.

Pratt, H. P. 1938. Ecology of the trout of the Gunnison River. Doctoral dissertation. Univ. Colorado.

Richardson, J. W., and A. R. Gaufin. 1971. Food habits of some Western stonefly nymphs. Trans. Am. Entomol. Soc. 97:91–122.

Short, R. A., and J. V. Ward. 1980. Life cycle and production of *Skwala parallela* (Frison) (Plecoptera: Perlodidae) in a Colorado mountain stream. Hydrobiologia 69(3): 273–275.

Stanford, J. A., and J. V. Ward. 1983. The effects of mainstream dams on physicochemistry of the Gunnison River Colorado. pp. 43–56. *In*: V. D. Adams and V. A. Lamarra, eds., Aquatic resources management of the Colorado River ecosystem. Ann Arbor Science Publishers, Ann Arbor, Mich.

Stanford, J. A., and J. V. Ward. 1983. The effects of regulation on the limnology of the Gunnison River: a North American case history. pp. 467–480. *In* A. Lillehammer and S. J. Saltveit, eds., Regulated rivers. Universitetforlaget, Oslo, Norway.

Ward, J. V. 1973. An ecological study of the South Platte River below Chessman Reservoir, Colorado, with special reference to macroinvertebrate populations as a function of the distance from the reservoir. Master's thesis. Univ. Colorado.

Ward, J. V. 1975. Downstream fate of zooplankton from a hydrolimnial release mountain reservoir. Int. Ver. Limnol. Verh. 19: 1798–1804.

Ward, J. V. 1976a. Comparative limnology of differentially regulated sections of a Colorado mountain river. Arch. Hydrobiol. 78(3): 319–342.

Ward, J. V. 1976b. Lumbricid earthworm populations in a Colorado, USA mountain river. Southwest. Nat. 21(1): 71–78.

Ward, J. V., and R. G. Dufford. 1979. Longitudinal and seasonal distribution of macroinvertebrates and epilithic algae in a Colorado springbrook pond system. Arch. Hydrobiol. 86(3): 284–321.

Woodling, J. 1975. The upper Gunnison River drainage. Water Quality Control Division, Colorado Department of Health, Denver, Colo.

Wu, Shi-Kuei. 1978. The fingernail and pill clams (family Sphaeriidae). *In* Natural history inventory of Colorado, vol. 2. Univ. Colorado Museum, Boulder, Colo. 60 pp.

CHAPTER 3

Tributaries in the Gila River System

INTRODUCTION

The Gila River and its tributaries are sometimes referred to as the Sonoran Desert streams. Gila River system streams are typical of arid area streams because they are characterized by large watershed areas that amplify the limited precipitation inputs, which form destructive flash floods. Flooding in desert streams is much more destructive than it is in mesic, temperate streams. Seasonally intermittent surface flow, high evaporation, and substantial subsurface flow are additional physical attributes. These streams are hard, alkaline, and somewhat saline. Because of sufficient nutrient levels and high light levels, these streams support extensive algal production and are largely autotrophic. The biota, which is greatly reduced by the destructive flooding, recovers rapidly.

These streams differ from most headwater streams in other parts of the United States in that they are classed as autotrophic rather than heterotrophic. They are heterotrophic only in pools.

Two well-studied and relatively undisturbed streams belonging to the Gila River system have been selected for this chapter. Aravaipa Creek, a tributary to the San Pedro River, is located in Graham and Pinal Counties. Sycamore Creek is a tributary of the Verde River and is located in Maricopa County. Both streams are part of the Gila River drainage in central Arizona.

PHYSICAL CHARACTERISTICS

Aravaipa Creek

Aravaipa Creek is an unregulated middle- to low-elevation perennial stream with a watershed area of 1400 km^2. The topography of the watershed is rugged and varied, with a maximum elevation of 3350 m above sea level. The stream channel begins at an elevation of 1400 m and ends at 625 m, where the Aravaipa meets the intermittent San Pedro River (Minckley, 1981).

The stream can be divided into physiographic sections. The upper section, Aravaipa Valley, is broadly incised (about 500 m deep and 10 km wide), and the stream channel has a floodplain of up to 2 km wide. Within this valley, perennial flow originates at 1010 to 1080 m above mean sea level from unconsolidated alluvium. From here it flows for 10 km, with gradients ranging from 2.5 to 10 m/km, until it reaches the middle section, Aravaipa Canyon. In this canyon, with walls sometimes exceeding 200 m, the floodplain usually does not exceed 100 m, and in some reaches the canyon bottom is only 30 m wide. The stream flows through this canyon for 17 km at gradients ranging from 5 to 25 m km^{-1}. Below the canyon, the stream travels through a progressively widening steep-sided valley until it meets the floodplain section of the San Pedro. Surface flow from the Aravaipa usually reaches the San Pedro only during flood stages. Most permanent tributaries are located in the canyon section, where they are typically short and precipitous (Minckley, 1981).

Geologically, Tertiary and Cretaceous volcanics comprise the majority of the Galiuro Mountains, which form the southern portion of the watershed. The Pinaleño and Turnbull–Santa Teresa Mountains, forming the northern and northeastern portion of the watershed, consist of Tertiary granites and a mixture of sedimentary, igneous, and metamorphic rocks. The Aravaipa Valley is underlain by Tertiary conglomerate and alluvium and Quaternary alluvium (Minckley, 1981).

The upper part of the Aravaipa Watershed is vegetated by desert grassland, while the lower creek flows through Sonoran Desert scrub. Mature riparian forest of cottonwood (*Populus fremontii*), willows (*Salix goodingii, S. bonplandiana*), sycamore (*Plantanus wrightii*), velvet ash (*Fraxinus pennsylvanica velutina*), box elder (*Acer negundo*), and wild grape (*Vitis arizonica*) are found in the upper valley. This community grades into one dominated by willows, cottonwood, and seepwillow (*Baccharis salicifolia*) downstream (Minckley, 1981).

Dense, fully leafed sycamores could reduce light levels to 20% of that present at midday during the summer, but light levels were generally reduced to only 50 to 70% in cottonwood and willow riparian zones. The shade from north-facing cliffs, such as those found in the canyon, can reduce summer light levels by 95% (Minckley, 1981).

Stream Channel. Natural stream pools, uncommon in Aravaipa Creek, occur next to cliffs and adjacent to obstructions. They rarely exceed 2 m in width and 0.5 m in depth (Barber and Minckley, 1966). Most of the bottom substrate of streams consists of riffles and runs separated by variously braided channels flowing over sand and gravel. Stream widths generally ranged from 4 to 10 m, and depths rarely exceeded 15 cm at modal discharge (Bruns and Minckley, 1980). Within the canyon, the stream typically flows as a single channel, which is narrower, deeper, and faster.

Rapids are relatively common in the canyon. Outside the canyon, the channel is wider, shallower, and often braided; pools are shallower, but more common, and rapids are uncommon (Minckley, 1981). The sediment in Aravaipa Creek consists of unconsolidated and poorly sorted sand, gravel, pebbles, cobbles, and boulders (Minckley, 1981).

Precipitation usually peaks bimodally, in winter (February and March) and summer (August). Summer rains are localized moderate to severe thunderstorms, while winter rains are more gentle, often spanning one or more days. The annual precipitation is 32 cm (Minckley, 1981).

The average annual discharge for lower Aravaipa Creek is 2530 h·m (hectaremeter) per year. Discharge patterns are extremely variable. For 90% of the time, the stream flow is less than 0.57 $m^3 s^{-1}$, accounting for about 40% of the total discharge. Destructive flooding (>2.83 $m^3 s^{-1}$) occurs less than 3% of the time but accounts for an additional 40% of the total discharge. The unconsolidated sediments of the creek bed and the varying depth of bedrock result in substantial longitudinal variation in surface discharge. Subsurface flow, particularly in the upper valley, accounts for a significant portion of the total discharge (Minckley, 1981).

As with precipitation, discharge tends to peak bimodally in winter and summer. Daily discharge is extremely variable, but during the winter, changes tend to be more gradual, often extending over several days. In dry winters (e.g., 1976), individual rainstorms have little direct effect on discharge. But in wet winters (e.g., 1978), discharges can reach flood conditions when rains occur for two or more consecutive days or for several alternating days within the same week. Summer storms correlate well with discharge, but individual storms rarely result in floods; summer floods are typically composites of inputs from several localities. Summer floods usually have higher intensities (i.e., the transition from modal to flood discharge is more sudden) and are therefore more destructive (Minckley, 1981) (Figure 3.1).

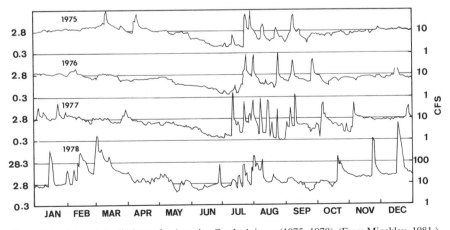

Figure 3.1.   Mean daily discharge for Aravaipa Creek, Arizona (1975–1978). (From Minckley, 1981.)

Floods originating in the upper valley carry large volumes of silt, sand, and gravel. This subjects the channel to alluviation or *sanding in*, creating smooth chute- or run-like conditions. Floods originating in the canyon tributaries have low sediment loads and higher kinetic energy. These degrade the channel and restore the riffle–run–pool conditions (Minckley, 1981).

When discharge is below 1.0 m^3 s^{-1} for at least 10 days, the turbidity is less than 1.0 JTU (Jackson turbidity unit). After the settlement of larger particles (three to five days), summer floodwater samples have readings of 14 to 56 JTU, while comparable winter floodwater samples consistently have values of over 100 JTU without agitation, indicating more semicolloids and colloids in the winter samples. The seasonal differences in floodwater turbidity reflect seasonal differences in water origin; winter floodwater comes primarily from the upper valley (Minckley, 1981).

Desert stream temperatures are subject to daily variations that are considerably greater than the usual maximum daily variation of 6.0°C for most small streams. The mean daily variation for Aravaipa Creek from June through August is about 9.6°C, while the December through February mean is about 6.7°C (Minckley, 1981).

Summer temperatures in Aravaipa Creek are coolest in the heavily shaded canyon. Water in the upper valley is cooler in summer and warmer in the other seasons than below the canyon, presumably because of greater subsurface flow and groundwater inflow in the upper valley (Table 3.1). The mean January stream temperature for the upper Aravaipa is 12.5°C, with a range from 10.5 to 14.5°C. In the lower Aravaipa, the January mean temperature is 8.0°C, with a range of 4 to 11.5°C (Minckley, 1981).

Sycamore Creek

Sycamore Creek has a watershed area of 505 km^2, with altitudes ranging from 427 to 2164 m above mean sea level; the study sites have elevations ranging from 552 to 640 m. As with Aravaipa Creek, the terrain is mountainous and rugged. During modal discharge the stream widths of the study sites range from 4 to 6 m, and the depth ranges from 5 to 7 cm. The summer modal discharge is 0.03 m^3 s^{-1} and the mean modal current velocity is about 0.015 m s^{-1}. Sand and gravel are the predominant substrates (Busch and Fisher, 1981; Fisher et al. 1982; Fisher and Gray, 1983; Gray,

TABLE 3.1. *Mean Midday Stream Temperatures (°C) in Aravaipa Creek, Arizona* [a]

	Upper Valley	Within Canyon	Below Canyon
Sept. 1976	26.7	22.8	—
	(24.3–28.4)	(22.0–24.3)	
June 1977	22.9	19.4	26.0
	(21.0–25.0)	(19.0–27.5)	(25.0–27.5)
July 1977	25.0	21.3	30.1
	(24.5–25.5)	(20.5–22.3)	(29.5–32.0)

Source: After Minckley (1981).
[a] Ranges are in parentheses.

1981). Unlike Aravaipa Creek, Sycamore Creek has intermittent flow in some reaches and has less extensive riparian vegetation.

CHEMICAL CHARACTERISTICS

Aravaipa Creek

The water of Aravaipa Creek is hard to very hard and has relatively high concentrations of sodium (Table 3.2). The mean specific conductance for upstream and downstream sites were 447.8 and 437.9 μS cm^{-1} at 25°C, respectively. Calcium concentrations, probably as a result of photosynthetic bicarbonate scavenging, decreased slightly downstream, whereas sodium increased downstream (Minckley, 1981).

Mercury (Hg^{2+}) concentrations (averaging 5.34 μg L^{-1} and ranging from 0 to 75 μg L^{-1}) of the main stem exceeded U.S. Environmental Protection Agency water quality criteria for freshwater life (0.05 μg L^{-1}). Main-stem cadmium (Cd^{2+}) concentrations (averaging 24.72 μg L^{-1} and ranging from 1.3 to 134.0 μg L^{-1}) exceeded its respective water quality standard (47 μg L^{-1}).

Stream pH varies between 7.4 and 8.9; the lower reaches are more alkaline. The pH is lower in the summer and at night. Summer lows may be due to the relative increase of subsurface flow, while nighttime lows are due to the decline in photosynthesis (Minckley, 1981).

Nitrogen and orthophosphate-P concentrations are generally higher after floods and during the winter. There is a general trend for N as NO_3 to decrease from upstream to downstream. This trend was not apparent for orthophosphate. N-nitrates vary from 0.13 to 0.61 mg L^{-1} upstream and from 0.02 to 0.49 mg L^{-1} downstream. Orthophosphate-P varies from 0.05 to 0.20 mg L^{-1} (Table 3.3). Nitrogen appears to be the limiting nutrient (Minckley, 1981), although the amounts of nitrogen and phosphorus are usually sufficient to support algal blooms. The higher amounts of phosphorus could produce greater amounts of growth if more nitrogen were present. The N/P ratios that favor blue-green algae, particularly nitrogen-fixing species, are less than 10:1. Some of these ratios in Aravaipa Creek are in this range.

The downstream decline in NO_3-N in desert streams is due largely to autotrophic uptake (Fisher, 1986). However, significant biological NO_3-N uptake was demonstrated for sediment cores from Sycamore Creek. Denitrification is one possible cause

TABLE 3.2. *Dissolved Concentrations (mg L^{-1}) of the Major Ions of Aravaipa Creek, Arizona*

Ca^{2+}	45.0	HCO_3^-	229.2
Mg^{2+}	9.3	CO_3^{2-}	112.8
Na^+	25.4	SO_4^{2-}	27.6
K^+	3.2	Cl^+	7.5

Source: After Minckley (1981).

TABLE 3.3.   *Downstream Changes in Macronutrient Concentrations (mg L^{-1}) in Aravaipa Creek, Arizona*

Date	Source	100 m	500 m	1000 m	2500 m
			a. Nitrate-N		
Aug. 1976	0.47	0.37	—	0.24	0.09
Oct. 1976	0.51	0.21	0.20	0.18	0.09
Jan. 1977	0.74	0.63	0.40	0.39	0.38
June 1977	0.39	0.41	0.21	0.09	0.11
Oct. 1977[a]	0.94	0.88	0.83	0.72	0.73
Dec. 1977[a]	1.07	0.93	0.97	0.83	0.85
Jan. 1978[a]	1.31	0.94	0.91	0.93	0.95
			b. Orthophosphate-P		
Aug. 1976	0.07	0.06	0.07	0.05	0.06
Oct. 1976	0.11	0.09	0.09	0.08	0.08
Nov. 1976	0.11	0.09	0.10	0.09	0.12
Jan. 1977	0.16	0.16	0.14	0.16	0.14
June 1977	0.08	0.09	0.09	0.07	—
Oct. 1977[a]	0.17	0.16	0.17	0.21	0.17
Jan. 1978[a]	0.21	0.22	0.19	0.18	0.21

Source: After Minckley (1981).

[a] These samples were taken within 30 days of flooding.

for this uptake, but this is doubtful given the highly oxygenated status of the hyporheic zone (Grimm and Fisher, 1984).

The mechanism for the regulation of phosphorus is not well understood because even though autotrophic uptake is certainly occurring, phosphorus does not show the trend in longitudinal depletion evident with nitrogen. Phosphorus levels in southwestern streams are near the equilibrium concentration for calcium phosphate, so it is possible that phosphorus may simply be controlled by its solubility (Fisher, 1986).

In the upper Aravaipa, daytime dissolved oxygen is highest in the late winter (generally between 9.0 and 11.5 mg L^{-1}) and lowest in late spring and summer (7.0 to 8.0 mg L^{-1}); differences between day and night concentrations are greatest during the late winter. In the lower reaches, the season of daytime highs in oxygen concentration is extended to include from middle autumn through late winter (9.0 to 10.0 mg L^{-1}). Late spring and summer lows are generally between 6.5 and 8.0 mg L^{-1}. Here the differences between day and night dissolved oxygen concentrations are more regular. Supersaturation is common during times of high primary production. Flash floods can create depressions in dissolved oxygen that last several days (Minckley, 1981).

Sycamore Creek

Sycamore Creek has moderately hard calcium and magnesium bicarbonate water with conductance values of 500 μS cm^{-1} and a total alkalinity of 250 mg L^{-1} as CaCO$_3$. Summer pH varies from 7 to 8. Nitrate-N usually exceeds 0.10 mg L^{-1} and dissolved phosphorus is generally below 0.050 mg L^{-1} (Fisher and Gray, 1983). During the postflood recovery period the soluble reactive phosphorus ranged from 0.054 to 0.094 mg L^{-1} and nitrates ranged from 0.070 to 0.140 mg L^{-1} (Fisher et al., 1982).

Dissolved oxygen in the interstitial water of Sycamore Creek is generally slightly lower than in surface water, but anoxic conditions were never found. Even at 0.5 m depth outside the wetted stream perimeter, oxygen is present at 3.85 mg L^{-1} (Grimm and Fisher, 1984).

ECOSYSTEM FUNCTION AND DYNAMICS

Detritus

Standing crops of stream detritus are greatest after flooding. The mean values for Aravaipa Creek detritus (particulate organic matter > 203 μm) were 25.8 to 47.0 g m^{-2} immediately after floods and 1.7 to 7.1 g m^{-2} with 45 or more days of modal discharge (Minckley, 1981).

The seasonal distributions of detrital standing crops in Aravaipa Creek reflect the effects of summer flooding. For 1975, 1976, and 1977, the winter and spring levels were relatively low, whereas the summer levels were consistently high, and the autumn levels reflect the residual effects of the summer peaks. After the onset of sustained flooding from the summer of 1977 through the spring of 1978, the standing crops were elevated. Standing crops also tended to be higher downstream (Minckley, 1981) (Table 3.4).

The sustained flooding of 1977 to 1978 apparently resulted in the progressive decline of detritus. The July 1978 flood resulted in an average of 46.2 g m^{-2}. After four major floods, the December mean standing crop was 27.5 g m^{-2}, the postflood January mean was 21.4 g m^{-2}, in March (after major floods in February) the mean was 12.2 g m^{-2}, and the single downstream sample in April was 17.6 g m^{-2} (Minckley, 1981).

In Sycamore Creek, the detritus [all particulate organic matter (POM)] standing crops were 231.6 g m^{-2} within 15 days of summer flooding, 75.7 g m^{-2} for 16 to 45 days after flooding, and 56.9 g m^{-2} for 46 to 90 days after flooding (Gray, 1980; as cited in Minckley, 1981). But even after flooding, Sycamore Creek fine particulate organic matter (FPOM) did not drop below 50 g m^{-2} (Fisher and Gray, 1983).

Stream detritus is predominantly FPOM. In Sycamore Creek, 82 to 94% of all POM was smaller than 100 μm, and fragments greater than 1 mm never comprised more than 6% (Fisher et al., 1982). The more extensive riparian forest at Aravaipa Creek may contribute more coarse particulate organic matter (CPOM), but this has not been quantified.

The bulk of the flood-associated detritus is FPOM of terrestrial origin (Fisher et al., 1981; Minckley, 1981). Most of this allochthonous detritus comes from plant fragments that were terrestrially processed into FPOM and then transported via sheet flow into the streams during rainstorms (Minckley, 1981).

TABLE 3.4. *Seasonal Standing Crops of Finely Divided Detritus (>203 μm) as Ash-Free Dry Weight (g m^{-2}) in Aravaipa Creek, Arizona*

Season[a]	Upper Valley	Within Canyon	Below Canyon
Spring 1975	3.8	7.0	15.6
Summer	53.8	94.7	42.0
Autumn	7.5	19.4	26.4
Winter 1976	2.2	6.5	8.6
Spring	5.4	6.7	9.6
Summer	39.8	44.1	31.2
Autumn	11.2	24.8	31.2
Winter 1977	1.9	2.6	1.6
Spring[b]	4.3	—	1.2
Summer	42.4	—	50.0
Autumn	18.7	—	36.3
Winter 1978	11.1	—	22.4
Spring	—	—	17.6

Source: After Minckley (1981).
[a] Winter, January–March; spring, April–June; summer, July–September; autumn, October–December.
[b] Includes preflood July samples.

Much of the flood-related influx of allochthonous FPOM is incorporated into the hyporheic zone (Fisher, 1986). On average, only 0.3% of the (sandy) sediment dry weight consisted of organic matter, but because of the depth and aeration of these alluvial deposits, this can amount to a substantial pool of metabolizable organic matter, an estimated 23.5 kg of organic matter per cubic meter of interstitial water (Grimm and Fisher, 1984).

Leaf Fall

Riparian leaf fall usually occurs in December, but despite the relatively well developed riparian vegetation at Aravaipa Creek, no significant accumulations have been observed (Minckley, 1981). Much of the autumnal leaf fall in desert streams impinges on the band of dry creek bed separating the stream from the riparian vegetation. Of the accumulated leaf fall that does impinge in the stream, much is probably exported because desert stream channels typically have few retention devices (e.g., snags) (Fisher, 1986). Prior to their removal by the turn of the century, riparian marshes may have served as sites for the accumulation of organic matter (Fisher et al., 1982) and may also have moderated the hydrological regime (Minckley, 1981), thereby facilitating additional accumulations.

Within a few days of the flood, the detritus standing crop declines via downstream or lateral export or in situ assimilation. During this decline, diatoms and filamentous algae colonize the newly scoured substrate, and as they increase, the detritus derived from these autochthonous producers becomes the major source of FPOM (Fisher et al., 1982; Minckley, 1981).

PRIMARY PRODUCTION

Algal biomass is high in Aravaipa Creek during periods of modal discharge, usually exceeding those recorded for temperate streams. The mean biomass for epipelic algae were 0.21, 0.01, and 2.0 g m^{-2} of ash-free dry weight for the upper valley, canyon, and lower valley, respectively. For the epilithic assemblage, the means were 1.48, 0.41, and 1.62 g m^{-2}, respectively. For the *Cladophora*–epiphytic assemblages, the means were 15.3, 3.2, and 26.4 g m^{-2}, respectively. Increased shading accounted for the lower biomass of *Cladophora* and other algae in the canyon (Minckley, 1981). *Cladophora* also tended to be less common in reaches shaded by riparian vegetation (Bruns and Minckley, 1980).

Algal biomass in Aravaipa Creek was greatly reduced immediately after flooding. Epilithic algae never exceeded 0.5 g m^{-2} ash-free dry weight, and the *Cladophora* assemblages never exceeded 6.0 g m^{-2} within 15 days of a spate. A recovery of biomass is usually achieved by 45 days (Minckley, 1981).

The seasonality of algal biomass reflects the destructiveness of summer flooding. Summer levels were typically the lowest, whereas winter and spring levels were typically the highest. The sustained flooding from July 1977 through April 1978 also reduced algal biomass to levels below those typical in summer (Table 3.5).

TABLE 3.5. *Seasonal Biomass of Filamentous Algae and Encrusting Algae[a] as Ash-Free Dry Weight (gm^{-2}) in Aravaipa Creek, Arizona*

Season[b]	Upper Valley	Within Canyon	Below Canyon
Spring 1975	30.6	5.4	13.5
Summer	5.4	0.0	1.1
Autumn	3.2	Trace	29.1
Winter 1976	13.4	4.3	56.0
Spring	4.3	6.5	14.0
Summer	4.8 (0.2)[b]	0.7 (0.1)	5.1 (0.7)
Autumn	9.7 (2.0)	1.4 (1.2)	20.3 (3.0)
Winter 1977	31.6 (2.3)	7.0 (0.4)	41.2 (1.0)
Spring[c]	36.9 (1.3)	— (—)	47.6 (1.2)
Summer 1977–spring 1978	0.3 (<0.1)	— (<0.1)	0.1 (<0.1)

Source: After Minckley (1981).
[a]Encrusting algae (i.e., epipelic and epilithic) biomass are in parentheses.
[b]Winter, January–March; spring, April–June; summer, July–September; autumn, October–December.
[c]Includes preflood July samples.

Algal biomass in Sycamore Creek is also high. Summer (July) preflood epipelic patches dominated by diatoms had mean patch ash-free dry weights (including detritus) of 174 g m^{-2} and mean chlorophyll *a* concentrations of 59 mg m^2 (Busch and Fisher, 1981). Patches of *Cladophora* + diatom epiphytes had mean ash-free dry weight and chlorophyll *a* concentrations of 363 g m^{-2} and 324 mg m^{-2}, respectively (Busch and Fisher, 1981). In October, the ash-free dry weight and chlorophyll *a* concentrations recorded by Fisher et al. (1982) for the same patch type were 240 g m^{-2} and 143 mg m^{-2}, respectively. The ash-free dry weight and chlorophyll *a* concentrations for the blue-green algae patch were 170 g m^{-2} and 140 mg m^{-2}, respectively, and for the *Cladophora* + epiphyte + blue-green algae patch they were 310 g m^{-2} and 190 mg m^{-2}, respectively (Fisher et al., 1982).

The ratio of chlorophyll *a* to ash-free dry weight is roughly 1:1000, whereas the ratios in the literature are nearer 1:100 (Sumner and McIntire, 1982; Vollenweider, 1974). These samples may have included detritus, which would increase the ash-free dry weight.

The pattern of flood-induced suppression and subsequent recovery of algal biomass observed in Aravaipa Creek is also evident in Sycamore Creek (Fisher et al., 1982) (Figure 3.2).

Primary production in Sonoran desert streams is high. Based on a regression model developed in situ, the preflood July gross primary production (GPP) in Sycamore Creek was estimated to be 8.5 g O$_2$ m^{-2} per day, the community respiration (CR) was estimated to be 5.1 g O$_2$ m^{-2} per day, and the P/R (GPP/CR) ratio was 1.7. For the *Cladophora* + epiphyte patch, the GPP was 12.5 g O$_2$ m^{-2} per day, the CR was 12.5 g O$_2$ m^{-2} per day, and the P/R ratio was 2.3. The GPP for the epipelic patch was 4.4 g O$_2$ m^{-2} per day, the CR was 4.6 g O$_2$ m^{-2} per day, and the P/R ratio was 0.96 (Busch and Fisher, 1981).

Using the same regression model, Fisher et al. (1982) found that both GPP and CR increased asymptotically in Sycamore Creek after a summer flood. By day 30 postflood, GPP was 6.6 g O$_2$ m^{-2} per day. By day 5 postflood, GPP exceeded CR and P/R

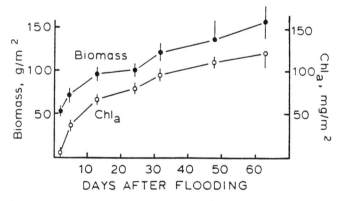

Figure 3.2.   Recovery of mean biomass (ash-free dry weight) and chlorophyll *a* concentrations following flooding in Sycamore Creek, Arizona. (From Fisher et al., 1982.)

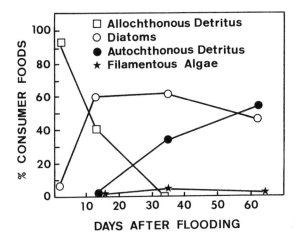

Figure 3.3. Proportion of foods in invertebrates following flooding. Data are based on nine species comprising 85% of the invertebrate biomass present and are weighted by the percent of total biomass represented by each of those species, Sycamore Creek, Arizona. (From Fisher et al., 1982.)

averaged 1.46 for the remaining two-month interval (Figure 3.3). Of the CR estimated by Fisher et al., about 22% is attributable to the insects. A large part of the rest is assignable to the plants, and probably only a small amount is assignable to fish (Fisher, 1986).

The P/R ratios for these Sycamore Creek studies indicate autotrophic conditions (i.e., P/R > 1.0) and a net exporting mode. The differences between net primary production and net biomass increment during the flood recovery period of Fisher et al. (1982) indicate that from day 13 to day 63 postflood, an average of 56% of the net production was exported while only 44% was utilized for the accumulation of new biomass. Some of this excess production was exported laterally, but most was exported downstream (Fisher et al., 1982).

A subsequent postflood study in Sycamore Creek revealed that (chamber) estimates of GPP and CR (14 and 25 days postflood) for the water column and surface sediment communities (the same communities measured in the previous studies) were autotrophic (P/R = 2.2). But with inclusion of hyporheic zone CR estimates (core samples), these estimates result in a P/R ratio that is approximately 1.0, indicating a balance between autotrophy and heterotrophy. The upstream–downstream O_2 method for estimating ecosystem metabolism resulted in higher estimates for both GPP and CR than those obtained for the chamber or chamber + core estimates, but the P/R ratios were similiar to those obtained with the chamber + core method (Grimm and Fisher, 1984) (Table 3.6).

On an aerial basis, the CR in the hyporheic zone (10 to 30 cm depth) is approximately equal to that of the surface zone. This metabolic activity is probably supported by the predominantly allochthonous dissolved organic matter and particulate organic matter that is incorporated into the sandy sediments during flooding. The high oxygen demand in the deeper sediments apparently does not result in anoxic conditions

TABLE 3.6.   *Gross Primary Production (g O_2 m^{-2} per day), Community Respiration*
(g O_2 m^{-2} per day), Production/Respiration Ratios, and Net Primary Production
(g O_2 m^{-2} per day) Estimates for Surface (Chamber), Combined Surface–Hyporheic
Communities (Chamber and Core), and Whole System (Upstream–Downstream
Differences in O_2) in Sycamore Creek, Arizona

	Chamber	Chamber + Core	Upstream–Downstream
Gross primary production			
Aug. 6 (14 days postflood)	7.3	7.3	9.4
Aug. 17 (25 days postflood)	8.0	8.0	10.3
Mean	7.6	7.6	9.85
Community respiration			
Aug. 6	3.1	7.8	10.1
Aug. 17	4.3	7.8	10.9
Mean	3.7	7.8	10.55
Production to respiration			
Aug. 6	2.4	0.9	0.93
Aug. 17	1.9	1.0	0.94
Mean	2.15	0.95	0.935
Net primary production			
Aug. 6	4.2	−0.5	−0.7
Aug. 17	3.7	0.2	−0.6
Mean	3.95	−0.15	−0.65

Source: After Grimm and Fisher (1984).

because of a thorough exchange between surface and interstitial waters (Grimm and Fisher, 1984).

The extent of metabolic coupling between the surface and autotrophic communities is not known, but the continual exchange of water between the surface and the sediment suggests that it may be significant. If the P/R ratio of Sycamore Creek does approach 1.0, Sycamore Creek is, by definition, not autotrophic. In any case, Sonoran streams such as Sycamore Creek are clearly heavily dependent on autochthonous production. Fisher (1986) suggested the possibility that the export of autotrophic production estimated in Fisher et al. (1982) may be offset by flood-imported allochthonous FPOM entrained in the sediment.

This possible autotrophic–heterotrophic relationship may be compromised by the method used to estimate export by Fisher et al. (1982). This method was based on difference calculations between net primary production (NPP) (approximately 2.0 g/m^2 per day) and net biomass increment. The estimates of NPP by Grimm and Fisher (1984), −0.7 to 0.2 g O_2 m^{-2} per day (Table 3.6), indicate that downstream export (as calculated by this method) is minimal.

Regardless of the difficulties involved in estimating the relationships between allochthonous and autochthonous inputs and exports, biomass does increase throughout the three-month postflood period, although the rate of increase appears to level off. With longer periods without destructive floods, biomass would probably continue to accumulate and vascular hydrophytes may become established. The exceptions would be when the primary producers become nutrient limited or when the stream dries (Fisher et al., 1982).

In Sycamore Creek, the degeneration of filamentous green algae has been consistently associated with low nitrate levels. This situation was often accompanied by an increase in nitrogen-fixing blue-green algae (e.g., *Anabaena variabilis*) (Fisher et al., 1982).

When the stream dries, the algae are stranded and the mobile invertebrates and fish are concentrated. In this case, respiration exceeds primary production, and the community is eventually reduced to fish and piscivorous Belastomatidae, which are carnivores or omnivores (Hemiptera) (Fisher et al., 1982).

SECONDARY PRODUCTION

In Aravaipa Creek, invertebrates are more numerous and have greater biomass in the upper valley, while the lowest mean density and biomass occur in the canyon. Density usually ranged from 1×10^5 to 2×10^5 individuals m^{-2}, and biomass generally ranges from 10 to 20 g m^{-2} (wet weight). Invertebrate abundance is strongly associated with the abundance of filamentous algae, detritus, and a coarse sand substrate that supports a well-diversified interstitial fauna (Bruns and Minckley, 1980; Minckley, 1981).

Flooding dramatically affects invertebrate populations in Aravaipa Creek. Densities were very low, often less than 50 individuals m^{-2} immediately after flooding. Yet by 15 days densities recovered to an average 4000 m^{-2} within and below the canyon and 8000 m^{-2} in the upper valley. The wet weight biomass was often reduced to less than 2 g m^{-2} immediately after flooding. Relative stability in density, biomass, and taxonomic richness was achieved by 46 to 90 days postflood (Minckley, 1981).

Baetis sp., *Callibaetis* sp., *Tricorythodes* sp. (Ephemeroptera), Chironomidae, and Simuliidae were abundant year-round. Oligochaeta, Acarina, and Ostracoda were numerous, and Hydropsychidae (Trichoptera) were major contributors to the biomass. *Ephemerella micheneri* (Ephemeroptera) was abundant during the spring, and *Mesocapnia frisoni* (Plecoptera) was abundant during the winter (Minckley, 1981).

Seasonal patterns of invertebrate abundance reflect the effects of summer flooding. Summer densities for most of the major taxa were typically lower than those of winter, spring, and autumn. Simuliidae were an exception. For both 1975 and 1976, they were more abundant during or after summer flooding (Minckley, 1981), probably because of the elevated flood-associated detritus levels (Bruns and Minckley, 1980) (Table 3.7).

Many short-lived invertebrates (e.g., Ephemeroptera and Chironomidae) usually recover rapidly from floods, but extended periods of rainfall and flooding (e.g., the winter of 1978) can inhibit their recovery. The sustained flooding from July 1977 through April 1978 in Aravaipa Creek virtually eliminated the (longer-lived) Trichoptera (Table 3.7) and greatly reduced populations of Odonata, Hemiptera, and Coleoptera (Minckley, 1981). The annual scenario of the operation of the system is

TABLE 3.7. *Mean Seasonal Density (Number/m^2) of Selected Benthic Insects in Aravaipa Creek, Arizona*

Season[a]	Ephemeroptera	Trichoptera	Chironomidae	Simuliidae
Winter 1975	974	33	2595	118
Spring	6762	326	3896	115
Summer	1974	204	562	822
Autumn	5704	496	4177	1598
Winter 1976	6828	711	4798	976
Spring	8555	472	2506	178
Summer	368	98	55	1111
Autumn	737	0	1586	309
Winter 1977	3908	184	1298	800
Spring	4042	646	2149	26
Summer	102	239	17	3
Autumn	1362	1	1946	942
Winter 1978	158	0	166	25
Spring	26	0	18	0

Source: After Minckley (1981).

[a] Winter, January–March; spring, April–June; summer, July–September; autumn, October–December.

that emergence occurs coincident with algal buildup in late winter and spring. Summer floods partially decimate the fauna, but density and standing crop continue to increase through autumn, accompanied by high detritus inputs and increases in algal growth with emergence occurring in late autumn. The *Cladophora* mats are important as food sources and habitats for the invertebrates (Bruns and Minckley, 1980).

The flooding regime in Sonoran desert streams probably selects for short-lived, small-bodied invertebrates. Compared to more temperate streams, their density tends to be higher relative to their biomass (Bruns and Minckley, 1980).

In Sycamore Creek, the mean invertebrate (dry weight) biomass is 3 g m^{-2}, with a range from near 0 to 10 g m^{-2}; the higher value reflects the autumn postflood maxima. The summer invertebrate fauna is dominated by collector–gatherers (87% of the density and 85% of the biomass). Of the collectors, short-lived Chironomidae and Ephemeroptera are the most numerous. These taxa, with their very short generation times (10 to 13 days), had a mean residence time (biomass/daily secondary production) of about 3 days, while the total collector fauna had a mean residence time of 4.7 days. Despite the modest biomass, the rapid turnover (annual production/biomass) ratio (45) results in a high annual production of 135 g m^{-2} per year (Fisher and Gray, 1983).

The secondary production of the collector component of macroinvertebrate communities in Sycamore Creek is high compared to other stream ecosystems. The estimate of

135 g m^{-2} per year (Fisher and Gray, 1983) is exceeded only by an estimate at Speed River in Ontario, where sediments of unconsolidated glacial till extended the invertebrate habitat nearly 1 m into the sediment. Noncollectors make up 15% of the invertebrate standing stock in Sycamore Creek. Most of these are larger and have longer life cycles than the collectors (Gray, 1981).

Secondary production in Sycamore Creek is dominated by a few taxa from the Ephemeroptera (*Baetis quilleri, Tricorythodes dimorphus, Leptohyphes packeri*) and Diptera (*Cryptolabis* sp., Chironomidae). Consequently, their species-specific annual production rates are also high. As an example, the annual production of *B. quilleri* is about tenfold higher (21.9 g m^{-2} per year) than for other species of *Baetis* in cooler temperate streams (0.9 to 2.1 g m^{-2} per year) (Fisher and Gray, 1983).

Sycamore Creek invertebrate biomass and density were reduced by summer flooding to less than 4% of their preflood levels. It takes about 13 days for the biomass to reach 50% of the preflood levels, while it takes about 35 days for the number of individuals to reach 50% recovery (Fisher et al., 1982). Trichoptera were extirpated from Sycamore Creek by extensive flooding from November 1978 to April 1979 (Gray, 1981).

Based on laboratory studies of the individual energy budgets of selected Sycamore Creek collectors and a mean standing stock of 3.0 g m^{-2}, the daily collector secondary production is 0.72 g m^{-2} per day, the respiration is 0.66 g m^{-2} per day, and the ingestion rate is 12.6 g m^{-2} per day. The ratio of ingestion to biomass (12.6/3.0 = 4.2) indicates that collectors ingest 4.2 times their body mass per day (Fisher and Gray, 1983). Considering that the summer gross primary production is 8.5 g m^{-2} per day (Busch and Fisher, 1981), Sycamore Creek collectors ingest about 1.5 times the primary production. During a two-month postflood recovery period, collectors ingested organic matter equivalent to 2.5 to 6.2 times the postflood primary production (Fisher and Gray, 1983).

Sycamore Creek invertebrates are estimated to turn over total organic matter every two to three days, much faster than the 71 to 102 days estimated for an English chalk stream. Since primary production by the dominant *Cladophora glomerata* is resistant to processing until it becomes detritus, other fractions of the organic matter pool must be processed at even higher rates (Fisher and Gray, 1983).

The extensive organic matter processing by Sycamore Creek invertebrates occurs because of high ingestion rates (44 to 251 μg mg^{-1} per hour) and exceedingly low assimilation efficiencies (7 to 15%). As an example, *Baetis quilleri* has a mean ingestion rate of 125 μg mg^{-1} per hour and an assimilation efficiency of 9.4%, whereas the Idaho mayfly, *Tricorythodes minutus*, has 24 μg mg^{-1} per hour and 27.5%, respectively (Fisher and Gray, 1983).

The rapid growth of Sycamore Creek invertebrates combined with high ingestion rates and low assimilation efficiencies is comparable to metabolic strategies adopted by organisms in temporary habitats. In desert streams, this metabolic strategy may be a response to abundant but low-quality food (e.g., *Cladophora* detritus). The strategy may be to shorten gut retention time, absorb only the most labile fraction, egest fecal material for microbial conditioning, and reingest this microbe–fecal complex (Fisher, 1986; Fisher and Gray, 1983).

FISH ABUNDANCE

As with the invertebrates, in Aravaipa Creek the mean fish density was highest in the upper valley (2.56 individuals m^{-2}), lowest within the canyon (0.72 individual m^{-2}), and intermediate below the canyon (1.27 individual m^{-2}). Density in the upper valley is low in winter, spring, and summer, but high in autumn. Density within the canyon was very low in the spring and higher in summer and autumn. Below the canyon, density peaked in summer and declined progressively toward spring (Minckley, 1981) (Table 3.8).

Both the canyon and the upper valley had similar (21.0 g m^{-2}) annual mean fish biomass, but their seasonal patterns were different. The lower valley had a mean of 14.4 g m^{-2}. There were seasonal peaks during the winter above and below the canyon, while the canyon peaked in summer. The upper valley was low during the summer, while the canyon was low during spring and autumn (Minckley, 1981) (Table 3.9).

Tagging experiments confirmed that at least some of the changes in fish abundance are attributable to seasonal dispersal. Fish tended to move toward the canyon during summer, and away from it during the winter, presumably as a response to differing temperature regimes (Minckley, 1981).

The large-bodied Sonoran sucker (*Catostomus insignis*) and the moderately large-bodied mountain sucker (*Pantosteus clarki*) dominated the biomass within and below the canyon, and came to dominate it in the upper valley during the spring and especially the winter. *P. clarki* was usually much more numerous than *C. insignis*. The small-bodied longfin dace (*Agosia chrysogaster*) and speckled dace (*Rhinichthys osculus*) numerically dominated the upper valley, except during the winter, when *P. clarki* was dominant. *A. chrysogaster* and *R. osculus* also came to dominate the upper valley biomass in summer and autumn. Except for the upper valley during winter, *A. chrysogaster* is the most numerous fish above and below the canyon. Within the canyon it was numerically dominant in summer and autumn, but it was rare there in spring and winter. *R. osculus* was numerically codominant with *A. chrysogaster* in the upper valley during summer and autumn, but was much less common in the canyon

TABLE 3.8.   *Mean (1976–1978) Seasonal Density (Number/m²) for Fish in Aravaipa Creek, Arizona*

Season [a]	Upper Valley	Within Canyon	Below Canyon
Winter	1.91	1.03	1.00
Spring	1.35	0.13	0.79
Summer	1.01	1.21	1.82
Autumn	5.99	0.53	1.46
Annual mean [b]	2.56	0.72	1.27

Source: After Minckley (1981).
[a] Winter, January–March; spring, April–June; summer, July–September; autumn, October–December.
[b] Annual means were calculated from the seasonal means.

TABLE 3.9.   *Mean (1976–1978) Seasonal Biomass (g m^{-2}) for Fish in Aravaipa Creek, Arizona*

Season[a]	Upper Valley	Within Canyon	Below Canyon
Winter	65.5	18.9	21.3
Spring	13.3	3.4	7.4
Summer	2.0	26.8	11.1
Autumn	10.8	6.4	14.9
Annual mean[b]	21.0	21.0	13.8

Source: After Minckley (1981).
[a] Winter, January–March; spring, April–June; summer, July–September; autumn, October–December.
[b] Annual means were calculated from the individual sample means.

and rare below it (Minckley, 1981). In Sycamore Creek, *A. chrysogaster* was consistently abundant (Fisher and Gray, 1983).

Flooding has less direct effect upon fish populations than it has upon algae and invertebrates. Changes in the relative abundance for the dominant Aravaipa Creek fish were not clearly related to variations in discharge. But variations in flooding appear to have had some effect on the populations of some of the other fish. The spikedace (*Meda fulgida*) increased in relative abundance during or after years of high water yield, while the roundtail chub (*Gila robusta*) and the loach minnow (*Tiaroga cobitis*) decreased in relative abundance (Minckley, 1981).

COMMUNITIES OF AQUATIC LIFE

Functional Relationships

Desert streams do not have large accumulations of allochthonous CPOM (i.e., leaf litter) (Fisher, 1984; Minckley, 1981). Consequently, the invertebrate detritivorous shredder fauna is represented primarily by only two genera. *Brillia* sp. (Chironomidae) was common in Sycamore Creek during the summer (Gray and Fisher, 1981) and may have been common in other seasons, and *Mesocapnia* spp. (Plecoptera) were collected only in the winter.

Mesocapnia arizonensis was collected only in Sycamore Creek during the winter. Nymphs first became abundant in the higher tributaries with the onset of winter rains and later became abundant at lower main-channel sites. It probably undergoes egg diapause for most of the year (Gray, 1981). *M. (= Capnia) frisoni*, abundant in winter, was present in Aravaipa Creek from autumn through early spring, and contributed about 3% to the total invertebrate density. It may also undergo egg diapause during the warmer months (Minckley, 1981). The maximum abundance for both of these stoneflies coincides with the occurrence of riparian leaf fall.

Graptocorixa serrulata (Hemiptera) and *Ochrotrochia* sp. (Trichoptera: Hydroptilidae) were the common macroalgal herbivores collected during the postflood recovery period in Sycamore Creek, but they accounted for only 1% of the total invertebrate density and biomass (Fisher et al., 1982). Hydrophilidae (Trichoptera), some of which may be detritivores, were common in Aravaipa Creek. *Graptocorixa* sp. and Limnephilidae (Trichoptera), some of which may also be herbivores, were much less so (Minckley, 1981). Several herbivorous miners and shredders of the Chironomidae (which are also collector–gatherers) were present in Sycamore Creek. At least one of them, *Cricotopus* sp., was common during the summer (Gray and Fisher, 1981) and autumn (Fisher et al., 1982).

FPOM detritivores and detritivore–herbivores (i.e., collector–gatherers and scrapers) dominated the invertebrate faunas of both Aravaipa and Sycamore Creeks. Collector–gatherers accounted for 87% of the density and 85% of the biomass in Sycamore Creek during the late summer and early autumn postflood period. *Cryptolabis* sp. (Diptera: Tipulidae), *Baetis quilleri*, *Leptohyphes packeri* (Ephemeroptera), Chironomidae, and Oligochaeta were the most common (Fisher et al., 1982). Collector–gatherers and scrapers contributed about 76% to the total density in Aravaipa Creek. *Baetis* spp., *Chloroterpes* sp., *Ephemerella* sp., *Tricorythodes* sp. (Ephemeroptera), Chironomidae, Glossomatidae, *Helicopsyche* sp. (Trichoptera), and Oligochaeta were the most common (Minckley, 1981).

The flood-associated succession of allochthonous detritus, diatoms, and autochthonous detritus is reflected in the diet of Sycamore Creek collectors. Gut contents reveal the dominance of allochthonous detritus in the collector's diet for the first 10 days after the flood. As diatom biomass increases, diatoms come to dominate the diet. Several weeks later, autochthonous detritus becomes about as important as diatoms to their diet. Filamentous algae is not an important diet component (Fisher et al., 1982) (Figure 3.4). The dominant filamentous algae *Cladophora glomerata* contains lauric acid, which is known to be harmful to many invertebrates.

Omnivorous collector–filter feeders were second in terms of density at Aravaipa Creek, contributing about 12%. Simuliidae and Hydropsychidae were abundant (Minckley, 1981). Filter feeders were recorded for Sycamore Creek (Gray, 1981) but were not collected in the postflood collections of Fisher et al. (1982) and were relatively uncommon in the postflood collections of Gray and Fisher (1981).

Invertebrate carnivores accounted for 13% of the density and 10% of the biomass during the postflood period in Sycamore Creek. Tanypodinae (Chironomidae), *Probezzia* sp. (Diptera: Ceraptogonidae), and Dytiscidae (Coleoptera) were common (Fisher et al., 1982). Carnivores accounted for about 7% of the total density at Aravaipa Creek. Chironomidae, Ceratopogonidae, and Dysticidae were the common carnivores (Minckley, 1981).

Among the fish in Aravaipa Creek, *Pantosteus clarki* is herbivorous, *Agosia chrysogaster* is omnivorous, and the remaining fish are carnivores. In addition to feeding on larger insects, *Gila robusta* also feeds on smaller fish, whereas the other carnivores feed only on invertebrates (Schreiber and Minckley, 1981).

The channel-dwelling *Pantosteus clarki* feeds predominantly on *Oedogonium* and diatoms; the intake of invertebrates is probably incidental. When *Cladophora* is abundant, *P. clarki* will selectively remove the epiphytes. After flood-induced scour, it will scrape diatoms from stones. *Agosia chrysogaster* feeds on *Mougeotia* sp., *Spirogyra* sp., and diatoms that are characteristic of stream margins and quiet

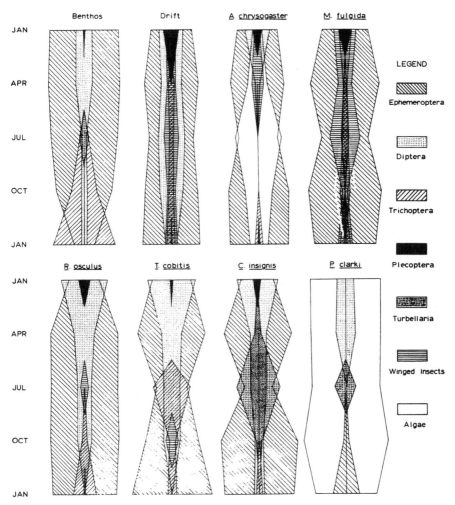

Figure 3.4. Relative abundance (by weight) of major invertebrate taxa in benthos and drift, and of invertebrate taxa and algae in stomachs (by volume) of fishes from Aravaipa Creek, Arizona, 1975 to 1976. (Reprinted by permission from "Feeding interrelationships of native fishes in a Sonoran Desert stream," by D. C. Schreiber and W. L. Minckley. Great Basin Naturalist 41: 417. © 1981, Great Basin Naturalist, Brigham Young University, Provo, UT 84602.)

waters. *Cladophora* contributes little to the diet of either of these fish (Schreiber and Minckley, 1981).

Except during the summer, there is considerable dietary overlap among the carnivores and omnivores, which are primarily fish. Ephemeroptera, predominant in the benthos and drift, are the most heavily utilized prey item. Of the mayflies, Baetidae are preyed upon most heavily; the differences in predation on other Ephemeroptera reflect the carnivore's habitat preferences and prey's seasonal availability. Seasonal peaks in Capniidae (Plecoptera) and Hydropsychidae (Trichoptera) are also reflected in fish diets. Preferential feeding of Diptera (mainly Chironomidae and Simuliidae) is evident for the bottom-feeding *Catostomus insignis*, *Rhinichthys osculus*, and *Tiaroga corbitis*, although the other fish also feed on Diptera (Schreiber and Minckley, 1981).

The feeding pattern diversifies and specializes in response to summer (and floods) and results in a decline in the density of the Diptera and in the declines in the density and body size of the Ephemeroptera. *Agosia chrysogaster* shifts toward eating algae, *Meda fulgida* increases its surface feeding on terrestrial and winged aquatic insects, *Tiaroga cobitis* selectively feeds on rheophilic Trichoptera, and *Catostomus insignis* selectively feeds on summer abundant Turbellaria. *Rhinichthys osculus* still prefers Ephemeroptera (Schreiber and Minckley, 1981).

There is a decrease in volume of food in fish stomachs during the summer, although it is not clear whether this is due to decreased consumption or to the increased gut evacuation rates that are usually associated with high temperatures (Schreiber and Minckley, 1981).

Both *Pantosteus clarki* and *Agosia chrysogaster* were herbivorous in Sycamore Creek. When food is plentiful, spatial and temporal food resource partitioning is evident for autumn populations of these two species. *A. chrysogaster* fed on and around mats of filamentous algae; *Cladophora glomerata*, *Zygnema*, epiphytic diatoms, and some epipelic diatoms were common gut contents. *P. clarki* probably fed in places where filamentous algae were less common. Epipelic diatoms are the common gut contents. *A. chrysogaster* fed during the day, while *P. clarki* had feeding peaks at dusk and dawn. During periods of apparent food scarcity, food partitioning between these two species was not evident (Fisher et al., 1981).

Primary Producers

In both Aravaipa and Sycamore Creeks, *Cladophora glomerata* is the most abundant filamentous algae. Other green algae, diatoms, and blue-green algae are usually found in epipelic assemblages or are associated with *C. glomerata* (Busch and Fisher, 1981; Minckley, 1981). Vascular hydrophytes are uncommon, probably because of the repeated flooding; isolated patches of *Nasturtium officinalis* have been noted in Aravaipa Creek (Barber and Minckley, 1966; Bruns and Minckley, 1980). There are historical records of marshes, or cienegas, in the upper valley of the Aravaipa, but these were largely destroyed by the turn of the century (Minckley, 1981).

Cladophora glomerata, *Mougeotia* sp., and *Spirogyra* sp. (green algae) are more common along the margins, while *Oedogonium* sp. is abundant in the main channel of Aravaipa Creek (Schreiber and Minckley, 1981). The epipelic and epilithic communities are dominated by diatoms (Minckley, 1981).

Under July preflood conditions, *Cladophora glomerata* occupied about 50% of the substrate at a Sycamore Creek study site. Its diatom epiphytes included *Cocconeis*, *Gomphonema*, *Navicula*, *Nitzschia*, and *Synedra*. Epipelic algae occupied the remaining substrate and was dominated by *Navicula* spp. and *Nitzschia* spp. (diatoms) and by *Nostoc* sp. and *Oscillatoria* (blue-green algae) (Busch and Fisher, 1981).

During May in Sycamore Creek, the differences among epilithic, epipelic, and epiphytic (on *Cladophora*) diatom patches were in terms of relative abundance rather than in terms of species composition. (The exception was *Cocconeis pediculus*, an epiphyte restricted to *Cladophora*.) Time, as well as substrate, affected patch differences. "Overgrown" (i.e., covered with brown flocculent material) epiphytic and epipelic patches were more like each other than they were with less overgrown epiphytic and epipelic patches. Within-patch diversity also declined in overgrown patches as *Nitzschia fonticola* gained in dominance (Busch, 1979).

In roughly decreasing order, the common diatoms were *Nitzschia fonticola* (30.4 to 60.4%), *Melosira varians* (4.5 to 15.1%), *Achnanthes lanceolata* (5.9 to 14.3%), *Navicula radiosa tenella* (2.2 to 11.2%), and *Gomphonema tenellum* (1.0 to 18.1%). Consistently dominant, *N. fonticola* became more so as a patch became "overgrown." *G. tenellum* was second in abundance on epilithic and "clean" epiphytic patches (Busch, 1979).

Shortly after summer flooding, bare sand in Sycamore Creek is rapidly colonized by an epipelic diatom patch dominated by *Achnanthes exigua*, *Gomphonema parvulum*, and *Navicula pupula*. Patches of blue-green algae (e.g., *Schizothrix calcicola* and *Anabaena variabilis*) and patches of *Cladophora* + cyanophytes + diatom epiphytes develop later in the flood-recovery sequence and come to be areally codominant with the epipelic patchs (Fisher et al., 1982).

Patches of *Cladophora* and epiphytic diatoms are largely restricted to the stream margins, where they are often stranded by peripheral drying. The epipelic patches are relatively persistent, but they have a 27% chance of being overgrown by blue-green algae. The blue-green patches have a similar chance (33%) of persistence, changing (via dislodgement) to the diatom patch (22%), or developing into the *Cladophora* + blue-green + epiphytic diatom patch (35%). The latter patch is the most persistent (71%) (Fisher et al., 1982).

Macroinvertebrates

Except for certain Hemiptera and adult Coleoptera (Gray and Fisher, 1981), few invertebrates are resistant to flooding, but many taxa are resilient (i.e., they recover quickly). In Aravaipa Creek (for the period from April 1975 to April 1976), Bruns and Minckley (1980) found that *Baetis* spp., *Chloroterpes inornata* (Ephemeroptera), Chironomidae (Diptera), and Oligochaeta were resilient, while *Tricorythodes* spp. (Ephemeroptera) and *Ochrotrichia* sp. (Trichoptera) were not. After an unusually intense year (1977) of flooding, Trichoptera were not found and Coleoptera, Hemiptera, and Odonata became very uncommon in Aravaipa Creek (Minckley, 1981). Trichopterans were also not collected from Sycamore Creek under similar conditions, and recolonization was probably effected from sources outside the watershed (Gray, 1981).

Based on Aravaipa Creek invertebrate densities (1975 to 1978) (presented in Minckley, 1981) the taxa that more quickly reestablish their populations are *Baetis* spp. (Ephemeroptera), Chironomidae, and Simuliidae. Numerically important taxa that appear to be vulnerable to summer flood suppression include *Chloroterpes* sp.,

Tricorythodes spp. (Ephemeroptera), Glossosomatidae (Trichoptera), Ceratopogonidae (Diptera), Acarina, Ostracoda, and Planaridae (Turbellaria). Numerically important taxa that appear to be vulnerable to both winter and summer flooding include the Hydropsychidae, Hydroptilidae (Trichoptera), and Oligochaeta. These vulnerable taxa, most particularly the Trichoptera, become very abundant in the absence of flooding.

Two other numerically important taxa, *Ephemerella micheneri* (Ephemeroptera) and *Mesocapnia frisoni* (Plecoptera), are seasonally restricted. *E. micheneri*, largely restricted to late winter and spring, appears to be vulnerable to extensive winter flooding. *M. frisoni*, restricted to late autumn, winter, and early spring, appears to be resistant to winter flooding.

Most of the Sycamore Creek Ephemeroptera, Diptera, and Hemiptera have development times and degree-day requirements that are among the lowest recorded for their respective orders. Reproduction is continuous and the number of generations per year for some of these taxa can exceed 35. This rapid development is a life-history trait that is shared with many insects from temporary aquatic habitats. These short development times increase the likelihood that there will be a sufficient reserve of aerial adults to recolonize after flooding (Gray, 1981).

Trichoptera reproduce year round but are usually more successful in autumn and spring, when flooding is unlikely. Because of higher stream temperatures, Sycamore Creek trichopterans have shorter generation times than do comparable taxa from temperate streams, but they have similar degree-day requirements. Based on these degree-day requirements, they can have up to seven generations per year. Sycamore Creek Coleoptera have development times that are comparable to those in more temperate streams, irrespective of any degree-day requirements. Unlike most insects in Sycamore Creek, most coleopterans exhibit strong seasonality. By concentrating their reproductive effort and subsequent larval development in either the spring and early summer or the late summer and early autumn, they usually avoid the worst effects of flooding. Most coleopterans have one or two generations per year (Gray, 1981).

Dormancy (i.e., aestivation or diapause) is not a commonly employed life-history strategy, probably because most sediments are either subject to unfavorable temperature regimes when dry or to extensive scour during flooding. *Mesocapnia* spp. (Plecoptera: Capniidae) and *Tabanus dorsifer* (Diptera: Tabanidae) are the only common taxa that undergo dormancy (Gray, 1981).

Longer-lived taxa usually avoid drought by selectively ovipositing in persistently aquatic habitats. Many Coleoptera will selectively oviposit in the deeper pools or main channels, while many Trichoptera will selectively oviposit in the upstream segments of cobble riffles. Such ovipositional habitat selection is not evident in the shorter-lived taxa (Gray, 1981).

Adult aerial dispersal is employed as the primary recolonization pathway by most stream insects. After frequent winter flooding in Sycamore Creek, most of the aerial colonists were nonreproductive adult Coleoptera and Hemiptera. After summer flooding, the aerial colonizers were dominated by reproductive Ephemeroptera and Diptera (Gray and Fisher, 1981).

Behavorial drift, particularly among immature Ephemeroptera and Chironomidae, enhances recolonization by dispersing insects from localized and aggregated ovipositional sites. Drift may also be an effective recolonization pathway in situations where the effects of flooding are localized. Upstream movement, predominantly by the same taxa that dominate the drift, is apparently stimulated during periods of elevated discharge. Postflood recolonization via vertical movement through the substrate has been demonstrated for some Ceratopogonidae, Chironomidae, and for *Tabanus dorsifer* (Diptera) (Gray and Fisher, 1981).

The coarser sediments of riffles are the primary habitat of the Simuliidae, Trichoptera (Hydropsychidae, Glossosomatidae, and Helicopsychidae), and the more rheophilic Ephemeroptera (*Chloroterpes inornata, Ephemerella michneneri, Heptagenia* sp., and *Rhithrogenia* sp.) (Schreiber and Minckley, 1981). In addition, riffles are probably the major habitat for rheophilic Plecoptera (*Mesocapnia* spp.), Coleoptera (Elmidae and Dryopidae), and Chironomidae (*Brillia* sp., *Eukiefferiella* sp., and Pentaneurini spp.). Because they support extensive growths of filamentous algae, riffles are also populated with Oligochaeta, Acarina, *Graptocorixa* sp. (Hemiptera), Hydroptilidae (Trichoptera), and Ceratopogonidae (Diptera). *Baetis* spp., probably the most abundant insects in Sonoran streams, occupy a variety of riffle and nonriffle habitats where filamentous algae and detritus are present.

Tricorythodes spp. (Ephemeroptera) are common on the sandy bottoms of the main channels (Schreiber and Minckley, 1981). Other insects found in the sandy bottom of the main channel include *Helicopsyche mexicana, Ochrotrichia* sp. (Trichoptera), *Probezzia* sp. (Ceraptogonidae), *Brillia* sp., *Cricotopus* sp. (Chironomidae), and *Euparyphus* sp. (Diptera: Stratiomyidae) (Gray and Fisher, 1981).

One Hydrophilidae (*Enochrus carinatus*) and several Diptera (Ceratopogonidae, Chironomidae, Tabanidae, and Tipulidae) were found on or in a saturated submerged sandbar in Sycamore Creek. Of these, the Ceratopogonidae (*Dasyhelea* sp. and *Probezzia* sp.), Tabanidae (*Tabanus dorsifer*), and Tipulidae (*Cryptolabis* sp.) could be found at depths of greater than 10 cm. *Micropsectra* sp., *Tribelos* sp., and *Pentaneurini* sp. (Chironomidae) were found in the same sandbar, but only at depths of less than 10 cm, and more commonly at the surface (Gray and Fisher, 1981).

During early autumn, the more abundant taxa found in Sycamore Creek at a shallow, slow-water site with sand, gravel, and filamentous algae included Oligochaeta, *Baetis quilleri, Leptohyphes packeri* (Ephemeroptera), *Probezzia* sp. (Ceratopogonidae), *Corynocera* sp., *Dicrotendipes* sp., *Micropsectra* sp. (Chironomidae), and *Cryptolabis* sp. (Tipulidae) (Fisher et al., 1982).

Physa virgata (Gastropoda), Acarina, Ostracoda, Gomphidae (Odonata), *Callibaetis* sp. (Ephemoptera), *Graptocorixa* sp., Veliidae (Hemiptera), Dytiscidae, Gyrinidae, Haliplidae, Hydrophilidae (Coleoptera), Hydroptilidae (Trichoptera), and Ceratopogonidae are also probably present in the slower shallows.

Polycentropus halidus, a filter-feeding trichopteran, can tolerate an absence of current. It has also been found in drying pools and stranded *Cladophora* mats in *Sycamore Creek* (Gray, 1981).

Throughout much of the year most tributaries of the Aravaipa are dry, except for isolated rockpools called tinajas. The more common tinajas macroinvertebrates are

Dytiscidae, Hydrophilidae (Coleoptera), Hemiptera, Odonata, and assorted Chirono-midae. Members of this fauna are usually uncommon in the main channel (Minckley, 1981). A comparable fauna is present in the side pools of the Sycamore (Gray, 1980). During the winter, as surface flow is renewed, these tributaries are colonized by main-stem insects. Thus these winter tributaries serve as sources for recolonization while the main stem is subjected to scour (Minckley, 1981).

Fish

The native fishes of Aravaipa Creek include the longfin dace (*Agosia chrysogaster*), speckled dace (*Rhinichthys osculus*), mountain sucker (*Pantosteus clarki*), Sonoran sucker (*Catostomus insignis*), loach minnow (*Tiaroga cobitis*), spikedace (*Meda fulgida*), and the roundtail chub (*Gila robusta*). Unlike in most natural Arizona waters, introduced species are not well established. Only the green sunfish (*Lepomis cyanellus*) is able to build populations, usually in the lower reaches, but it is consistently suppressed by flooding (Minckley, 1981).

In Aravaipa Creek, these seven natives have been persistent (i.e., consistently present) and have had a stable assembledge (i.e., there has been little change in their relative abundance for almost 40 years) (Meffe and Minckley, 1987). They have behavorial and morphological adaptations that help them to maintain their position during flooding. On the other hand, introduced species have evolved under lowland mesic conditions, where downstream displacement or lateral escape onto flood-plains are viable responses to flooding. Under desert conditions these responses usually result in stranding (Minckley and Meffe, 1987).

Agosia chrysogaster is adapted for life in small, hot lowland Sonoran streams. It is extremely persistent in drying pools and is tolerant of high temperatures, and it is particularly effective in colonizing intermittent reaches (Barber and Minckley, 1966; Deacon and Minckley, 1974). It is widespread and abundant in Aravaipa Creek but is seasonally uncommon within the canyon (Minckley, 1981). It is the only consistently abundant fish in the intermittent Sycamore Creek (Fisher and Gray, 1983)

The youngest *Agosia chrysogaster* are most common along the banks and in shallow eddies, while large adults are prevalent in pools, often seeking cover under algal mats (Barber and Minckley, 1966). The most common habitats are sand-bottomed runs with moderate currents; swift and still waters appear to be avoided (Minckley, 1981; Schreiber and Minckley, 1981).

Rhinichthys osculus is near the lower end of its altitudinal range in Aravaipa Creek; a fact reflected in its abundance in the upper valley and scarcity elsewhere. It has some habitat overlap with *Agosia chrysogaster*, but it is most common in riffles. Large adults are often found in deeper runs and the flowing parts of pools, usually with some kind of cover (Minckley, 1981; Schreiber and Minckley, 1981).

Pantosteus clarki and *Catostomus insignis* are common to abundant and wide-spread in Aravaipa Creek. Very young *P. clarki* live in quiet waters near the banks, small adults and young are concentrated in riffles and rapids, and larger adults inhabit strongly flowing, deeper water, usually near cover (Barber and Minckley, 1966; Schreiber and Minckley, 1981). The larger adults will move into faster water to feed (Minckley, 1981). *C. insignis* primarily inhabits pools, seeking cover during the day and moving out to the shallows, or even the riffles, to feed at night (Minckley, 1981). Both of these suckers are regarded as common inhabitants of larger streams and rivers (Minckley, 1981).

Tiaroga cobitis and *Meda fulgida*, both Gila River endemics, are uncommon in Aravaipa Creek. *T. cobitis* inhabits gravel and cobble riffles, typically in association with beds of *Cladophora*. Its populations, probably as a consequence of this association, are enhanced somewhat when the flooding frequency is low (Minckley, 1981). *M. fulgida* inhabits moving, laminar waters, usually at a depth of less than 30 cm. Concentrations are often encountered at the downstream ends of riffles or in eddies near the upper ends of pools. Turbulence is avoided (Barber and Minckley, 1966; Minckley, 1981; Schreiber and Minckley, 1981).

Populations of *Gila robusta* have probably been decimated with the "filling in" of pools associated with watershed overgrazing during the last century. In Aravaipa Creek, this uncommon fish is largely restricted to the canyon reach, where it inhabits small eddies behind side canyons and occasional pools and pockets beneath ledges (Barber and Minckley, 1966). It is secretive and tends to cluster in the deepest, most permanent pools. It is known to move into swift water, presumably to feed (Minckley, 1981).

For *Rhinichthys osculus* and *Pantosteus clarki*, recruitment occurs in winter, while it usually occurs in spring or early summer for the other fishes. *Agosia chrysogaster* spawns throughout the year but is usually most successful in spring, early summer, and again in late summer after the cessation of flooding (Minckley, 1981).

Agosia chrysogaster, *Gila robusta*, and *Meda fulgida* select pools and runs for spawning. *A. chrysogaster* will burrow its eggs into the barren silt or sand of shallow waters, while *Gila robusta* will broadcast its eggs over the sand and silt of the deeper pools, and *M. fulgida* will broadcast over silt, sand, and gravel in shallow water. *Tiaroga cobitis* will attach its eggs in a nest on the undersides of cobbles in riffles. *Rhinichthys osculus* will place its eggs within the substrate of shallow riffles. *Pantosteus clarki* and *Catostomus insignis* will place their eggs within the gravelly substrate of shallow to moderately deep rapids, riffles, and runs (Minckley, 1981).

Functional Ecosystems of Aquatic Life in Aravaipa Creek

Vegetative Habitats. The vegetative habitats were those found among the plants floating or attached and slime that was algae, probably blue-greens growing attached to sediment particles in the bottom of pools or slow-flowing areas. Those organisms found associated with plants were the green algae *Coelastrum* sp., an unidentified species of *Oedogonium*, and various unidentified species of the genus *Nitzschia* (Tables 3.10 and 3.11).

Crawling over the plants were the omnivorous species of *Physa*; unidentified species of Oligochaete, which are probably omnivores; and unidentified species of spiders. Other animals found in this habitat were Crustacea, an omnivorous unidentified species belonging to the genus *Cypris*. Also in and among the vegetation were several odonates belonging to the genera *Coenagrion* and *Ischnura*. These are probably carnivores. Also present was a carnivorous odonate, *Aeshna* sp. A few mayflies were found in association with the vegetation. They were unidentified species of

TABLE 3.10. *Structure of Functioning Ecosystem: Aravaipa Creek, Vegetative Habitats*

Carnivores

Abedus herberti	Graptocorixa sp.	Peltodytes sp.
Aeshna sp.	Gyrinus sp.	Plea sp.
Ambrysus sp.	Ischnura spp.	Probezzia sp.
Cataclysta sp.	Mesovelia sp.	Ranatra sp.
Coenagrion spp.		

Omnivores

Callibaetis sp.
Cypris sp.
Unidentified Oligochaete
Physa sp.
Sigara sp.

Detritivores

Detritivore–herbivores

Detritivore–omnivores
Baetis sp.
Mesocapnia (Capnia) frisoni

Herbivores

Herbivores
Agosia chrysogaster
Berosus sp.
Laccobius sp.
Pantosteus clarkii

Herbivore–omnivores
Hydrophilids
Ochrotricha spp.

Detritus

Algae

Associated with plants:
Coelastrum sp.
Nitzschia spp.
Oedogonium sp.

In slime:
Chlorococcum sp.
Hantzshia sp.

TABLE 3.11. *Species List: Aravaipa Creek*

Taxon[a]	Vg	Fm	Sl	Sb	Hs	Source[c]	Feeding Habits[d]
SUPERKINGDOM PROKARYOTAE							
KINGDOM MONERA							
Division Cyanophyta							
Class Cyanophyceae							
Order Nostocales							
Family Nostocaceae							
Anabaena sp.						Mi1	
Nostoc sp.				×		F1,Mi1	
Family Oscillatoriaceae							
Lyngbya sp.		×				Mi1	
Oscillatoria sp.					×	Mi1	
Spirulina sp.						Mi1	
Family Rivulariaceae							
Gleocapsa sp.					×	Mi1	
SUPERKINGDOM EUKARYOTAE							
KINGDOM PLANTAE							
Subkingdom Thallobionta							
Division Chlorophycota							
Order Tetrasporales							
Family Tetrasporaceae							
Tetraspora sp.		×			×	Mi1	
Order Volvocales							
Family Chlamydomonadaceae							
Chlamydomonas sp.				×		Mi1	
Order Chlorococcales							
Family Chlorococcaceae							
Chlorococcum sp.			×			Mi1	
Family Coelastraceae							
Coelastrum sp.	×					Mi1	
Family Hydrodictyaceae							
Hydrodictyon sp.						Mi1	
Pediastrum sp.		×				Mi1	
Family Oocystaceae							
Ankistrodesmus sp.		×				Mi1	
Family Scenedesmaceae							
Scenedesmus sp.						Mi1	
Order Ulotrichales							
Family Cylindrocapsaceae							
Cylindrocapsa sp.						Mi1	
Family Microsporaceae							
Microspora sp.		×			×	Mi1	
Family Trentepohliaceae							
Gongrosira sp.					×	Mi1	
Family Ulotrichaceae							
Ulothrix sp.		×			×	Mi1	
Order Chaetophorales							
Family Chaetophoraceae							
Stigeoclonium sp.					×	Mi1	
Order Oedogoniales							
Family Oedogoniaceae							
Oedogonium sp.	×	×				Mi1,S	
Order Siphonocladales							
Family Cladophoraceae							

(continues)

TABLE 3.11. *Continued*

Taxon[a]	Vg	Fm	Sl	Sb	Hs	Source[c]	Feeding Habits[d]
			Habitat[b]				
Cladophora glomerata					×	Ba,Br, Bu1, Bu2, F1,F3, G3, Mi1	
Rhizoclonium sp.					×	Mi1	
Order Zygnematales							
Family Zygnemataceae							
Mougeotia sp.		×				Mi1	
Spirogyra sp.		×				Mi1	
Zygnema sp.					×	F1,Mi1	
Division Chromophycota							
Class Bacillariophyceae							
Order Eupodiscales							
Family Coscinodiscaceae							
Cyclotella sp.						Mi1	
Melosira sp.						Mi1	
Order Fragilariales							
Family Fragilariaceae							
Diatoma spp.						Mi1	
Fragilaria spp.						Mi1	
Meridion spp.						Mi1	
Synedra spp.						Mi1	
Order Achnanthales							
Family Achnanthaceae							
Achnanthes spp.						Mi1	
Cocconeis spp.						Mi1	
Order Naviculales							
Family Cymbellaceae							
Amphora spp.						Mi1	
Caloneis sp.						Mi1	
Cymbella spp.						Mi1	
Family Gomphonemaceae							
Gomphonema spp.						Mi1	
Family Naviculaceae							
Navicula spp.						Mi1	
Pinnularia spp.						Mi1	
Order Epithemales							
Family Epithemiales							
Denticula spp.			×			Mi1	
Hantzschia sp.						Mi1	
Rhopalodia sp.						Mi1	
Order Bacillariales							
Family Nitzschiaceae							
Nitzschia spp.	×			×		F3,Mi1	
Order Surirellales							
Family Surirellaceae							
Surirella spp.						Mi1	
Class Xanthophyceae							
Order Vaucheriales							
Family Vaucheriaceae							

TABLE 3.11. *Continued*

Taxon[a]	Habitat[b]					Source[c]	Feeding Habits[d]
	Vg	Fm	Sl	Sb	Hs		
Vaucheria spp.						Mi1	
Class Charophyceae							
Order Charales							
Family Characeae							
Chara sp.					×	Mi1	
Division Magnoliophyta							
Class Dicotyledoneae							
Order Capparales							
Family Brassicaceae							
Nasturtium officinale				×		Ba,Bu1	

Taxon	Habitat[e]								Source[c]	Feeding Habits[d]
	Ms	Sc	T	D	V	S	G	C		
KINGDOM ANIMALIA										
Subkingdom Protozoa										
Class Mastigophora										
Order Euglenoidida										
Family Euglenidae										
Euglena sp.						×			Mi1	O
Phacus sp.						×			Mi1	O
Phylum Cnideria										
Class Hydrozoa										
Family Hydridae										
Hydra sp.	×								Bu1,Mi1	C
Phylum Platyhelminthes										
Class Turbellaria										
Order Tricladida										
Family Planariidae										
unident. spp.	×	×		×		×	×		Mi1,S	C
Phylum Nemata										
unident. spp.	×	×	×	×		×	×		Bu1,Mi1	
Phylum Nematomorpha										
Class Gordioida										
unident. spp.	×	×							Mi1	C
Phylum Mollusca										
Class Gastropoda										
Order Basommatophora										
Family Physidae										
Physa sp.	×	×	×		×	×	×	×	Mi1	C
Class Bivalvia										
Order Heterodonta										
Family Sphaeriidae										
Pisidium sp.		×				×			Mi1	D
Phylum Annelida										
Class Hirudinoidia										
unident. spp.	×	×							Bu1,G2,Mi1	C
Class Oligochaeta										
unident. spp.	×	×	×	×	×	×	×		Bu1,Mi1	O
Phylum Arthropoda										
Class Arachnida										
Order Acariformes										

(continues)

TABLE 3.11. *Continued*

	Ms	Sc	T	D	V	S	G	C	Source[c]	Feeding Habits[d]
				Habitat[e]						
unident. spp.	×	×		×	×	×			G2,Mi1	
Class Crustacea										
Subclass Cephalocarida										
Order Cladocera										
unident. spp.		×	×						Mi1	D
Subclass Ostracoda										
unident. spp.			×	×		×			Mi1	O
Order Podocopina										
Family Cypridae										
Cypris sp.	×	×		×		×			Mi1	O
unident. spp.	×			×		×			Bu1	O
Subclass Copepoda										
unident. spp.			×						Mi1	O
Class Insecta										
Order Odonata										
Suborder Zygoptera										
Family Coenagrionidae										
Argia sp.	×			×			×	×	Bu1,G2	C
Coenagrion sp.	×	×			×				Mi1	C
Ischnura sp.	×	×			×				Mi1	C
Suborder Anisoptera										
Family Aeshnidae										
Aeshna sp.			×		×				Mi1	C
Family Gomphidae										
Aphylla sp.	×	×				×	×		Mi1	C
Erpetogomphus sp.	×					×			Bu1	C
Ophiogomphus sp.	×					×			Bu1	C
unident. spp.	×	×	×						Mi1	C
Family Libellulidae										
Dythemis sp.	×	×	×			×			Mi1	C
unident. spp.		×	×						Bu1,Mi1	C
Order Ephemeroptera										
Suborder Shistonota										
Family Baetidae										
Baetis spp.	×	×		×	×	×	×	×	Bu1,Mi1, S	DO
Callibaetis sp.	×	×	×		×				Bu1,Mi1	O
Centroptilum spp.	×								Bu1,G2, Mi1	DO
Family Heptageniidae										
Heptagenia sp.		×					×	×	Mi1	O
Rhithrogena sp.	×	×					×	×	Mi1,S	OD
Family Leptophlebiidae										
Chloroterpes inornatus	×	×					×	×	Mi1,S	O
Suborder Pannota										
Family Ephemerellidae										
Ephemerella micheneri	×	×				×	×	×	Bu1,Mi1	O
Family Tricorythidae										
Leptohyphes apache	×								Mi1	O
Tricorythodes condylus	×					×			Bu1,Mi1, S	O
T. dimorphus	×					×			Bu1,G2, G3,Mi1, S	O

TABLE 3.11. *Continued*

	Habitat[e]								Source[c]	Feeding Habits[d]
	Ms	Sc	T	D	V	S	G	C		
T. minutus	×	×			×				Bu1,Mi1, S	O
Order Plecoptera										
Family Capniidae										
Mesocapnia (= *Capnia*) *frisoni*	×	×			×		×	×	Bu1,Mi1, S	OD
Order Hemiptera										
Family Belostomatidae										
Abedus herberti	×	×	×	×		×			Bu1,G2, G3,Mi1	C
Lethocerus medius				×					Bu1,Mi1	C
Family Corixidae										
Graptocorixa sp.	×	×			×				Bu1,Mi1	C
Sigara sp.		×			×				Mi1	O
Family Gelastocoridae										
Gelastocoris sp.	×	×				×	×		Mi1	C
Family Gerridae										
Gerris sp.		×	×						Mi1	C
Family Macroveliidae										
unident. spp.		×							Mi1	C
Family Mesoveliidae										
Mesovelia sp.			×		×				Bu1,Mi1	C
Family Naucoridae										
Ambrysus sp.	×	×			×	×	×		Bu1,Mi1	C
Family Nepidae										
Ranatra sp.		×			×				Bu1,G2, Mi1	C
Family Notonectidae										
Buenoa sp.			×						Mi1	C
Notonecta sp.		×	×						Mi1	C
Family Pleidae										
Plea sp.		×			×				Mi1	C
Family Veliidae										
Microvelia sp.	×	×							Mi1	C
Rhagovelia sp.		×							Mi1	C
Order Megaloptera										
Family Corydalidae										
Corydalus sp.	×			×		×	×	×	Bu1,Mi1	C
Order Coleoptera										
Suborder Adephaga										
Family Dytiscidae										
Bidessus sp.	×		×						G2,G3, Mi1	C
Deronectes sp.	×	×	×		×				Bu1,Mi1	C
Hydrophorus spp.	×	×			×				G2,Mi1	C
Laccophilus sp.			×						Mi1	C
Thermonectus sp.	×		×		×				Bu1,Mi1	C
unident. spp.		×	×						Mi1	C
Family Gyrinidae										
Gyrinus spp.		×	×	×	×				G2,Mi1	C
Family Haliplidae										
Peltodytes spp.	×	×	×		×				Bu1,G2, Mi1	C
unident. spp.			×						Mi1	C
Suborder Polyphaga										
Family Dryopidae										

(continues)

TABLE 3.11. *Continued*

	Habitat [e]								Source [c]	Feeding Habits [d]
	Ms	Sc	T	D	V	S	G	C		
Helichus sp.	×	×					×	×	Bu1,Mi1	O
Family Elmidae										
unident. spp.	×	×							Bu1,Mi1	O
Family Helodidae										
unident. spp.		×							Mi1	O
Family Hydrophilidae										
Berosus spp.		×	×		×				Mi1	HO
Hydrobius sp.	×								Bu1,G2	O
Laccobius spp.	×		×		×				G2,Mi1	H
Tropisternus sp.	×	×	×						Bu1,Mi1	O
Family Psephenidae										
Psephenus sp.	×						×		Mi1	O
Order Trichoptera										
Family Glossosomatidae										
unident. spp.	×						×	×	Bu1,Mi1	O
Family Helicopsyche										
Helicopsyche sp.	×					×	×	×	Bu1,Mi1	O
Family Hydropsychidae										
Cheumatopsyche sp.	×	×					×	×	Bu1,Mi1	O
Hydropsyche sp.	×	×					×	×	Bu1,Mi1	O
Family Hydroptilidae										
Leucotricha sp.	×								Bu1,Mi1	O
Mayatrichia sp.	×								Bu,Mi1	O
Ochrotricha spp.	×	×			×	×			Bu1,G2,Mi1	HO
Family Limnephilidae										
unident. spp.	×	×							Bu1,Mi1	O
Family Rhyacophilidae										
Rhyacophila sp.		×					×	×	Mi1	OC
Order Lepidoptera										
Family Pyralidae										
Cataclysta sp.	×				×				Bu1,Mi1	H
Order Diptera										
Suborder Nematocera										
Family Blephariceridae										
Blepharicera sp.	×								Mi1	O
Family Ceratopogonidae										
Forcipomyia sp.		×	×						Mi1	O
Probezzia sp.	×	×			×	×			G2,G3,Mi1	C
unident. spp.	×	×							Bu1,Mi1	O
Family Chironomidae										
unident. Chironominae	×	×	×						Mi1	O
unident. Orthocladiinae	×	×							Mi1	O
unident. Pentaneurini	×	×	×			×			G2,Mi1	CO
unident. Tanypodini	×	×	×						Mi1	C
unident. Tanytarsini	×						×	×	Mi1	O
unident. spp.	×								Bu1	
Family Culicidae										
Aedes sp.			×						Mi1	O
Anopheles sp.	×	×							Mi1	O
Culex sp.			×						Mi1	O
unident. spp.	×								Bu1	O
Family Dixidae										

TABLE 3.11. *Continued*

	Ms	Sc	T	D	V	S	G	C	Source[c]	Feeding Habits[d]
				Habitat[e]						
Dixa sp.				×					Mi1	O
unident. spp.	×	×		×					Bu1	O
Family Psychodidae										
Pericoma sp.	×	×		×					Mi1	O
unident. spp.	×								Bu1	O
Family Simuliidae										
Simulium sp.	×	×				×	×	×	×	Mi1	O
unident. spp.	×								Bu1,Mi1	O
Family Tipulidae										
unident. spp.	×	×	×						Bu1,Mi1	OD
Suborder Brachycera										
Family Dolichopodidae										
unident. spp.	×								G2,Mi1	C
Family Empididae										
unident. spp.	×	×		×					Bu1,Mi1	CO
Family Ephydridae										
unident. spp.	×	×			×				Bu1,G2, Mi1	O
Family Muscidae										
unident. spp.	×	×	×						Bu1,G2, Mi1	C
Family Scatophagidae										
unident. spp.	×	×	×		×				Mi1	CH
Family Stratiomyidae										
unident. spp.	×	×							Bu1,Mi1	O
Family Tabanidae										
Silvius sp.	×								Mi1	C
Tabanus spp.	×	×	×						Mi1	C
unident. spp.	×								Bu1	C
Phylum Chordata										
Subphylum Vertebrata										
Class Osteichthyes										
Order Cypriniformes										
Family Catosomidae										
Catostomus insignis									Ba,Me, Mi1, Mi2,S	O
Pantosteus clarki									Ba,F1,Me, Mi1, Mi2,S	O
Family Cyprinidae										
Agosia chrysogaster									Ba,F1,Me, Mi1, Mi2,S	O
Gila robusta								×	Ba,Me, Mi1, Mi2,S	O
Meda fulgida									Ba,Me, Mi1, Mi2,S	O
Rhinichthys osculus									Ba,Me, Mi1, Mi2,S	O
Tiaroga cobitis									Ba,Me, Mi1, Mi2,S	O
Order Cypridontiformes										
Family Poeciliidae										

(continues)

TABLE 3.11. *Continued*

	Habitat [e]							Source [c]	Feeding Habits [d]
	Ms	Sc	T	D	V	S	G	C		
Gambusia affinis									Mi1,Mi2	O
Order Perciformes										
Family Centrarchidae										
Lepomis cyanellus									Mi1,Mi2	C

[a]An asterick indicates an introduced species.

[b]Autotroph habitats: Vg, vegetation (epiphytic); Fm, floating mat; Sl, slime; Sb, sediment benthos (epipelic); Hs, hard substrate (epilithic).

[c]Ba, Barber and Minckley (1966); Br, Bruns and Minckley (1980); Bu1, Busch (1979); Bu2, Busch and Fisher (1981); D, Deacon and Minckley (1974); F1, Fisher et al. (1981); F2, Fisher and Gray (1983); F3, Fisher et al. (1982); G1, Gray (1980); G2, Gray (1981); G3, Gray and Fisher (1981); Me, Meffe and Minckley (1987); Mi1, Minckley (1981); Mi2, Minckley and Meffe (1987); S, Schreiber and Minckley (1981).

[d]Feeding habits: C = carnivore; CH = carnivore–herbivore; CO = carnivore–omnivore; D = detritivore; DO = detritivore–omnivore; H = herbivore; HO = herbivore–omnivore; O = omnivore; OC = omnivore–carnivore; OD = omnivore–detritivore; OH = omnivore–herbivore.

[e]Invertebrate habitats: Ms, main stem; Sc, side canyon; T, tinajas; D, detritus; V, vegetation; S, sand; G, gravel; C, cobble (and boulder).

Baetis, which may be detritivore–omnivores, and an unidentified species of the genus *Callibaetis*, an omnivore. Other species in this habitat were unidentified species probably belonging to *Mesocapnia (Capnia) frisoni*, which is a detritivore–omnivore and the hemipteran *Abedus herberti*, a carnivore. A few corixids were found associated with the vegetation. They were the carnivorous *Graptocorixa* sp., and *Sigara* sp., which is probably an omnivore. Also present were an unidentified species of the genus *Mesovelia*, a carnivore, and *Ambrysus* sp., also a carnivore, as is *Ranatra* sp. and a species belonging to the genus *Plea*. Beetles were found associated with the vegetation. They belong to the genera *Gyrinus* and *Peltodytes*, which are carnivorous. Several hydrophillids were found associated with the vegetation, which were herbivore–omnivores. *Berosus* sp. and *Laccobius* sp. were herbivores. The trichopteran *Ochrotricha* was found in and among the vegetation. Several unidentified species of this genus were present. They are probably herbivore–omnivores. A lepidopteran, *Cataclysta* sp., was also found among the vegetation, as was the dipteran *Probezzia* sp., which is a carnivore. An unidentified species of blackflies was also found among the vegetation, as were unidentified species of the family Ephydridae and the family Scatophadidae.

Here and there were floating mats of plants and debris. Here and there in pools were beds of slimy consistency probably consisting of blue-green algae, although these were not identified. In this slime were found an unidentified species of the genus *Chlorococcum* and an unidentified species of *Hantzschia*.

Sandy Benthos: Sand and Gravel Habitats.   The benthos consisted of sand and mud and sometimes pebbles. Also associated with it was some plant material, but this was relatively rare. The main habitats were sand and mud and gravel. Here were found an unidentified species of the blue-green algae *Nostoc* and of a green alga, *Chlamydomonas* sp. Also present attached to the substrate were species of the genus *Nitzschia*. Growing in the substrates was a dicot, *Nasturtium officinale*. The proto- zoans were found in this habitat. They were the omnivorous *Euglena* sp. and *Phacus* sp.; unidentified species of flatworm, which are probably carnivores; and unidentified species belonging to the phylum Nemata. Also present were unidentified species of the genus *Physa*, which are probably omnivores, but some tend to be carnivorous. Also, unidentified Oligochaetes were found, which are probably omnivores. Ostra- cods, which were unidentified and are probably omnivores, were also present, as was an unidentified species of the genus *Cypris* sp. that is an omnivore. Most of the fam- ily Cypridae found in this area are probably omnivores (Tables 3.11 and 3.12).

Several odonates were in and among the sand and gravel. They were an unidentified species of the carnivorous genus *Argia*. Also present was an unidentified species of the genus *Aphylla*, which is a carnivore, and of the genera *Erpetogomphus* sp. and *Ophio- gomphus* sp., both of which are carnivores. Also present belonging to the family Li- bellulidae was an unidentified species of the genus *Dythemis*, a carnivore. Other mayflies present were unidentified species of the genus *Baetis*, detritivore–omnivores; an unidentified species of *Heptagenia*, an omnivore; and of the genus *Rhithrogena*, which is an omnivore–detritivore. The omnivorous *Choroterpes inornatus* was also present, as was the omnivorous mayfly, *Ephemerella micheneri*. Other mayflies that were found in this habitat that were omnivorous were *Tricorythodes condylus, T. dimor- phus,* and *T. minutus*. One species of Hemiptera, *Abedus herberti,* which is a carni- vore, was present in the sand and gravel and the vegetation associated with it. An unidentified species of *Gelastocoris,* which is a carnivore, was also present. This is, of course, a hemipteran. Other species of insects found in this habitat were an uniden- tified species of *Ambrysus,* which is a carnivore, and *Corydalus,* a carnivore. Several beetles were found in this habitat: the carnivorous *Deronectes,* the carnivorous *Hydro- phorus,* and the carnivorous *Thermonectus*. Also present were gyrinid beetles, which were unidentified to species. Also present was the omnivorous trichopteran *Helico- psyche* and unidentified species of *Cheumatopsyche* and *Hydropsyche,* both omni- vorous. Also present was the trichopteran species belonging to the genus *Ochrotricha,* which are herbivore–omnivores, and the omnivore–carnivore species *Rhyacophila*. The dipteran *Probezzia,* a carnivore, was also present, and an unidentified species of chironomid, probably belonging to the genus *Tanytarsini*. Species of this genus are omnivores. Also present was a carnivorous–omnivorous species which was unidenti- fied but belonged to *Pentaneurini*. Blackfly larvae were also found in this habitat. They are unidentified species of *Simulium,* which are omnivorous.

Lotic Habitats: Hard Substrates. Associated with the swifter water on hard substrates were the blue-green algae *Oscillatoria* sp. and *Gleocapsa* sp. The green algae present were specimens of the unidentified species of *Tetraspora* and of the Microsporaceae, *Microspora* sp., and specimens of the unidentified species of the genus *Gongrosira*. Other green algae found in the faster water associated with hard substrates were *Ulothrix* sp., *Stigeoclonium* sp., *Cladophora glomerata,* an unidenti- fied species of *Rhizoclonium,* and an unidentified species of *Zygnema*. Although there probably were many diatoms in this habitat, they were not identified. A few

TABLE 3.12. *Structure of Functioning Ecosystem: Aravaipa Creek, Sandy Benthos, Sand and Gravel Habitats*

Carnivores

Abedus herberti	Deronectes sp.	Hydrophorus sp.
Ambrysus sp.	Dythemis sp.	Ophiogomphus sp.
Aphylla sp.	Erpetogomphus sp.	Probezzia sp.
Argia sp.	Unidentified flatworms	Thermonectus sp.
Corydalus sp.	Gelastocoris sp.	

Omnivore–carnivores

Pentaneurini sp.
Physa sp.
Rhyacophila sp.

Omnivores

Cheumatopsyche sp.	Helicopsyche sp.	Physa spp.
Choroterpes inornatus	Heptagenia sp.	Simulium spp.
Most of the family Cypridae	Hydropsyche sp.	Tanytarsini sp.
found in this area	Unidentified Oligochaetes	Tricorythodes condylus
Cypris sp.	Unidentified Ostracods	T. dimorphus
Ephemerella micheneri	Phacus sp.	T. minutus
Euglena sp.		

Detritivores

Detritivore–herbivores	**Detritivore–omnivores**	**Herbivores**
	Baetis sp.	
	Rhithrogena sp.	

Detritus

Herbivores

Herbivores	**Herbivore–omnivores**
Chlamydomonas sp.	Ochrotricha spp.
Nasturtium officinale	

Algae

Nitzschia spp.
Nostoc sp.

specimens of *Chara* sp. were found associated with these hard habitats in faster water (Tables 3.11 and 3.13).

Several animals were found in this habitat. Crawling over the rocks was an unidentified species of snails belonging to the genus *Physa*, which are believed to be carnivorous. A few mayflies were found in this habitat. They were unidentified species of *Baetis*, which are probably detritivore–omnivores; *Heptagenia* sp., an omnivore; of *Rhithrogena*, an omnivore–detritivore; and specimens of the omnivorous *Chloroterpes inornatus* and of *Ephemerella micheneri*, which is an omnivore. Other animals found associated with the faster water and hard substrates was a megalopteran *Corydalus* sp., which is a carnivore, and the beetles *Helichus* sp., an omnivore found in and among the rocks. Also present was a hydrophilid, *Psephenus* sp., an omnivore; undetermined trichopterans; an undetermined species belonging to the genus *Helicopsyche*, an omnivore; and *Cheumatopsyche*, which is an omnivore, as is *Hydropsyche* sp. Other trichopterans living on the rocks in the faster-flowing water were an omnivore–carnivore *Rhyacophila* sp. A few chironomids were found in this habitat. They were probably omnivorous and belonged to the *Tanytarsini*. Blackflies belonging to the genus *Simulium*, which are omnivores, were found associated with the rocks in this faster-flowing water.

Functional Ecosystems of Aquatic Life in Sycamore Creek

Sycamore Creek is typical of streams in desert climates where the phosphates are high compared with most eastern streams. Similarly, metals such as calcium and magnesium are high in these types of streams. The associations of aquatic life were what one would expect in these types of streams.

Vegetative Habitats. Associated with vegetation were diatoms such as *Melosira varians*, *Fragilaria capucina*, *Meridion circulare*, and *Synedra ulna*. Other diatoms associated with the vegetation were two taxa belonging to the genus *Achnanthes* (*A. lanceolata* and *A. minutissima*) and two taxons belonging to the genus *Cocconeis*. The genus *Amphora* was represented by three taxons, and the genus *Cymbella* by two taxons. These taxons were probably growing attached to the vegetation, as were five taxons belonging to the genus *Gomphonema*. In among the vegetation where the flow of water was relatively minor were the diatoms *Navicula cuspidata*, *N. exigua capitata*, *N. minuscula*, *N. mutica*, *N. pupula*, *N. radiosa tenella*, and *N. viridula avenacea*. Growing attached to the vegetation were the species *Epithemia turgida* and *Hantzschia amphioxys*. Six taxons belonging to the genus *Nitzschia* were probably growing attached to the vegetation in regions where the current was relatively slow.

Several invertebrates were found associated with the vegetation. They were *Physa virgata*, which is a carnivore, sometimes with omnivorous habits. Also present were undetermined species of spider. Odonates were found in and among the vegetation. They were *Pantala hymenea*, which is a carnivore; and *Siphlonurus* sp., a mayfly, which is a carnivorous–omnivorous species sometimes acting as a carnivore, sometimes as an omnivore. Also present was an unidentified species of *Caenis*, an omnivore. Hemipterans were represented by the carnivorous *Abedus herberti*, three species from the family Corixidae and *Ranatra* sp. A few beetles were found in and among the vegetation: *Agabus seriatus* and five taxons belonging to the genus *Deronectes*,

TABLE 3.13. *Structure of Functioning Ecosystem: Aravaipa Creek, Lotic Habitats, Hard Substrates*

Carnivores
Corydalus sp.
Physa sp.

Omnivore–carnivores
Rhyacophila sp.

Omnivores
Cheumatopsyche sp.
Choroterpes inornatus
Ephemerella micheneri
Helichus sp.
Helicopsyche sp.
Heptagenia sp.
Hydropsyche sp.
Psephenus sp.
Simulium sp.
Tanytarsini spp.

Detritivore–omnivores
Baetis sp.
Rhithrogena sp.

Herbivores
Chara sp.
Cladophora glomerata
Gelocapsa sp.
Gongrosira sp.
Microspora sp.
Oscillatoria sp.

Herbivore–omnivores
Rhizoclonium sp.
Stigeoclonium sp.
Tetraspora sp.
Ulothrix sp.
Zygnema sp.

Detritivores

Detritivore–herbivores

Herbivores

Detritus

Algae

which are carnivorous. Other beetles that were present were the carnivorous *Dytiscus marginicollis* and unidentified species of the carnivorous genus *Hydrophorus*. Other species of Colleoptera found in and among the vegetation were *Oreodytes* sp., a carnivore, *Rhantus* sp., and *Thermonectes marmoratus*, which are carnivores. The gyrinids were represented by species in two genera, *Dineutus* sp. and *Gyrinus* spp., which are carnivorous. The herbivorous *Haliplus* sp. and the carnivore–herbivore species of the genus *Peltodytes* were also present. Other insects found in and among the vegetation were *Berosus exilis* and *B. punctatissimus*. Both of these species are herbivore–omnivores. Also present were the herbivorous *Laccobius* spp. Species of the genus *Ochrotricha*, which are Trichoptera, were present. They are believed to be herbivore–omnivores. Other species associated with the vegetation were the trichopteran *Polycentropus hallidus* and the lepidopteran *Paragyractus confusialis*, which are herbivores. A dipteran that was present was the carnivorous *Probezzia* sp. Other species present were the chironomids *Brillia* sp., which is an omnivore–detritivore; *Chironomus* sp., an omnivore–herbivore; and *Corynocera* sp., an omnivore as well as a species of *Polypedilum*, an omnivore. Other species found in and among the vegetation were unidentified species of the family Ephydridae, one of the flies.

Various algae, such as blue-greens and some of the greens, often formed a slimy surface on some of the harder substrates, such as twigs or logs or stones or pebbles. In this habitat were diatoms, *Cyclotella meneghiniana* and *Melosira varians*. Other diatoms found in this habitat were *Fragilaria vaucheriae, Meridion circulare, Synedra rumpens*, and *S. ulna*. A few *Eunotia curvata* were present in this habitat. The family Achnanthaceae was represented by four taxons, the genus *Amphora* by two taxons, the genus *Caloneis bacillum* by one taxon, and the genus *Cymbella* by three taxons. The genus *Gomphonema* was represented by four taxons (Tables 3.14 and 3.15). The genus *Navicula* was quite diverse in and among the "slime." Present were 11 taxons belonging to the genus *Navicula, Stauroneis anceps, Epithemia sorex*, and *Hantzschia amphioxys*. Other diatoms in and among the filaments of algae that formed the slimy substrate were seven taxons belonging to the genus *Nitzschia* (Tables 3.14 and 3.15) and *Surirella angustata*.

Sand Habitats and Sand and Gravel Habitats. Among the sediments, sometimes a mixture of sand and soil, mainly sand, was an ecosystem composed of several species of algae, invertebrates, and other organisms. Here one found a blue-green alga, *Anabaena variabilis*, and an unidentified species of *Nostoc*. Several diatoms were found in and among the sediments. They were *Cyclotella meneghiniana, Melosira varians, Fragilaria vaucheriae, Meridion circulare, Synedra ulna*. There were also three taxons belonging to the genus *Achnanthes* sp. in this habitat (Tables 3.15 and 3.16) and one taxon belonging to the genus *Cocconeis* (*Cocconeis placentula* var. *lineata*). Also present were *Amphora submontana, A. venata, Caloneis bacillum, Cymbella hustedtii*, and *C. minuta*. The genus *Gomphonema* was represented by four taxons: *G. angustatum, G. olivaceum, G. parvulum*, and *G. tenellum*. The genus *Navicula* was represented by 10 taxons in this habitat, that is, in and among the sediments (Tables 3.15 and 3.16). Also present in this habitat were *Pinnularia viridus* and *Epithemia sorex*. The genus *Nitzschia* was quite common in the muddy–sandy substrate. Seven taxons were found in this substrate (Tables 3.15 and 3.16). Also present was *Surirella angustata*.

TABLE 3.14. *Structure of Functioning Ecosystem: Sycamore Creek, Vegetative Habitats*

————————————————— Carnivores —————————————————

Abedus herberti	*D. nebulosus*	*D. yaquii*	*Gyrinus* spp.	*Pantala hymenea*
Agabus seriatus	*D. roffi*	*Dineutus* sp.	*Hydrophorus* spp.	*Rhantus* sp.
Deronectes aequinoctialis	*D. striatellus*	*Dytiscus marginocollis*	*Oreodytes* sp.	*Thermonectes marmoratus*

————————— Omnivore–carnivores —————————

Physa virgata
Siphlonurus sp.

————————— Omnivores —————————

Agosia chrysogaster
Caenis sp.
Corynocera sp.
Dasyhelea sp.
Polypedilum sp.

——————————————— Detritivores ———————————————

Detritivores	**Detritivore–omnivores**	**Herbivores**
	Brillia sp.	*Haliplus* sp.
		Laccobius spp.
		Pantosteus clarki
		Paragyractus confusialis
		Polycentropus hallidus

——————————————— Herbivores ———————————————

Herbivore–omnivores	**Herbivore–carnivores**
Berosus exilis	*Peltodytes* sp.
B. punctatissimus	
Chironomus sp.	
Ochrotricha spp.	

Detritus	Algae

Detritus

Achnanthes lanceolata
A. minutissima
Amphora perpusilla
A. submontana
A. veneta
Cocconeis pediculus
C. placentula var. lineata
Cymbella hustedtii
C. minuta
Epithemia turgida
Fragilaria capucina
Gomphonema angustatum
G. lanceolatum
G. olivaceum
G. parvulum
G. tenellum
Hantzschia amphioxys
Melosira varians
Meridion circulare
Navicula cuspidata
N. exigua capitata
N. minuscula
N. mutica
N. pupula
N. radiosa tenella
N. viridula avenacea
Nitzschia dissipata
N. fonticola
N. frustulum
N. linearis
N. palea
Nitzschia spp.

Algae

In slime:

Achnanthes lanceolata
A. minutissima
Amphora submontana
A. venata
Caloneis bacillum
Cyclotella meneghiniana
Cymbella hustedtii
C. minuta
C. sinuata
Epithemia sorex
Eunotia curvata
Fragilaria vaucheriae
Gomphonema angustatum
G. olivaceum
G. parvulum
G. tenellum
Hantzschia amphioxys
Melosira varians
Meridion circulare
Navicula atomus
N. cryptocephala
N. cuspidata
N. exigua capitata
N. minima
N. minuscula
N. mutica
N. pupula
N. radiosa tenella
N. secreta apiculata
N. viridula avenacea
Nitzschia acicularis
N. amphibia
N. dissipata
N. fonticola
N. frustulum
N. linearis
N. palea
Stauroneis anceps
Surirella angustata
Synedra rumpens
S. ulna

TABLE 3.15   *Species List: Sycamore Creek*

Taxon[a]	Vg	Fm	Sl	Sb	Hs	Source[c]	Feeding Habits[d]
SUPERKINGDOM PROKARYOTAE							
KINGDOM MONERA							
Division Cyanophyta							
Class Cyanophyceae							
Order Nostocales							
Family Nostocaceae							
Anabaena variabilis				×		F3	
Nostoc sp.				×		F1,Mi1	
Family Oscillatoriaceae							
Oscillatoria lutea					×	F3	
Schizothrix calcicola				×	×	F1	
Family Rivulariaceae							
Calothrix parietina					×	F1	
SUPERKINGDOM EUKARYOTAE							
KINGDOM PLANTAE							
Subkingdom Thallobionta							
Division Chlorophycota							
Division Chlorophyta							
Order Volvocales							
Family Chlamydomonadaceae							
Chlamydomonas globosus				×		F3	
Order Siphonocladales							
Family Cladophoraceae							
Cladophora glomerata					×	Ba,Br, Bu1, Bu2,F1, F3,G3, Mi1	
Order Zygnematales							
Family Zygnemataceae							
Zygnema sp.					×	F1,Mi1	
Division Chrysophyta							
Class Bacillariophyceae							
Order Eupodiscales							
Family Coscinodiscaceae							
Cyclotella meneghiniana			×	×	×	Bu1	
Melosira varians	×		×	×	×	Bu1,F1	
Order Fragilariales							
Family Fragilariaceae							
Fragilaria capucina	×					Bu1	
F. vaucheriae			×	×		Bu1	
Meridion circulare	×		×	×	×	Bu1	
Synedra rumpens			×			Bu1	
S. ulna	×		×	×	×	Bu1	
Order Eunotiales							
Family Eunotiaceae							
Eunotia curvata			×			Bu1	
Order Achnanthales							
Family Achnanthaceae							
Achnanthes exigua				×		F3	
A. lanceolata	×		×	×	×	Bu1	
A. minutissima	×		×	×	×	Bu1	
Cocconeis pediculus	×		×			Bu1,F1	

TABLE 3.15 *Continued*

Taxon[a]	Habitat[b]					Source[c]	Feeding Habits[d]
	Vg	Fm	Sl	Sb	Hs		
C. placentula						F1	
C. placentula var. *lineata*	×		×	×	×	Bu1	
Order Naviculales							
Family Cymbellaceae							
Amphora perpusilla	×					Bu1	
A. submontana	×		×	×	×	Bu1	
A. venata	×		×	×	×	Bu1	
Caloneis bacillum			×	×		Bu1	
Cymbella hustedtii	×		×	×	×	Bu1	
C. minuta	×		×	×	×	Bu1	
C. sinuata			×			Bu1	
Family Gomphonemaceae							
Gomphonema angustatum	×		×	×	×	Bu1	
G. lanceolatum	×				×	F1	
G. olivaceum	×		×	×	×	Bu1	
G. parvulum	×		×	×	×	Bu1,F3	
G. tenellum	×		×	×	×	Bu1	
Family Naviculaceae							
Navicula atomus			×	×		Bu1	
N. cryptocephala			×	×	×	Bu1	
N. cuspidata	×		×			Bu1	
N. exigua capitata	×		×	×	×	Bu1	
N. minima			×	×	×	Bu1	
N. minuscula	×		×	×	×	Bu1	
N. mutica	×		×	×	×	Bu1	
N. pupula	×		×	×	×	Bu1,F3	
N. radiosa tenella	×		×	×	×	Bu1,F1	
N. secreta apiculata			×	×	×	Bu1	
N. viridula						F1	
N. viridula avenacea	×		×	×	×	Bu1	
Pinnularia viridus				×		Bu1	
Stauroneis anceps			×			Bu1	
Order Epithemales							
Family Epithemiaceae							
Rhopalodia gibba					×	Bu1	
Family Epithemiales							
Epithemia sorex			×	×		Bu1	
E. turgida	×					Bu1	
Family Nitzschiaceae							
Hantzschia amphioxys	×		×		×	Bu1	
Order Bacillariales							
Family Nitzschiaceae							
Nitzschia acicularis			×			Bu1	
N. amphibia			×	×		Bu1	
N. dissipata	×		×	×	×	Bu1	
N. fonticola	×		×	×	×	Bu1	
N. frustulum	×		×	×	×	Bu1	
N. linearis	×		×	×	×	Bu1	
N. palea	×		×	×	×	Bu1	
Nitzschia spp.	×			×		F3,Mi1	
Order Surirellales							
Family Surirellaceae							
Surirella angustata	×		×	×	×	Bu1	
S. ovata					×	Bu1	

(continues)

TABLE 3.15   *Continued*

			Habitat[e]					Source[b]	Feeding Habits[c]
	Ms	Sc	T	D	V	S	G	C		
KINGDOM ANIMALIA										
Phylum Platyhelminthes										
Class Turbellaria										
Order Tricladida										
Family Planariidae										
Dugesia sp.									G2	C
Phylum Mollusca										
Class Gastropoda										
Order Basommatophora										
Family Physidae										
Physa virgata	×				×	×	×	×	F2,G2	C
Phylum Annelida										
Class Hirudinoidia										
unident. spp.	×	×							Bu1,G2, Mi1	C
Class Oligochaeta										
Order Haplotaxida										
Family Tubificidae										
unident. spp.	×					×			G2	O
Phylum Arthropoda										
Class Arachnida										
Order Acariformes										
unident. spp.	×	×		×	×	×			G2,Mi1	
Class Insecta										
Order Odonata										
Suborder Zygoptera										
Family Calopterygidae										
Hetaerina sp.				×					G2,G3	C
Family Coenagrionidae										
Argia sp.	×			×		×	×		Bu1,G2	C
Suborder Anisoptera										
Family Gomphidae										
Ophiogomphus bison									G2,G3	C
Progomphus borealis					×				G2,G3	C
Family Libellulidae										
Pantala hymenea					×	×			G2	C
Order Ephemeroptera										
Suborder Shistonota										
Family Baetidae										
Baetis insignificans									G2	DO
B. quilleri	×					×			F2,G2,	DO
Callibaetis montanus									G2	O
Centroptilum spp.	×								Bu1,G2, Mi1	DO
Family Leptophlebiidae										
Thraulodes speciosus						×	×		G2	DO
Family Siphlonuridae										
Siphlonurus sp.				×	×				G2	CO
Suborder Pannota										
Family Caenidae										

TABLE 3.15 *Continued*

	Ms	Sc	T	D	V	S	G	C	Source[b]	Feeding Habits[c]
				Habitat[e]						
Caenis sp.					×				G2	O
Family Tricorythidae										
Leptohyphes packeri	×					×			F2,G2, G3	O
Tricorythodes dimorphus	×					×			Bu1,G2, G3,Mi1, S	O
Order Plecoptera										
Family Capniidae										
Mesocapnia arizonensis	×		×				×	×	G2,G3	D
Family Taeniopterygidae										
unident. spp.									G2	DO
Order Hemiptera										
Family Belostomatidae										
Abedus herberti	×	×	×	×	×				Bu1,G2, G3,Mi1	C
Family Corixidae										
Graptocorixa serrulata	×		×		×				F3,G2,G3	C
Hesperocorixa sp.					×				G2	C
Trichocorixa reticulata					×				G2	C
Family Naucoridae										
Ambrysus occidentalis	×								G2,G3	C
Family Nepidae										
Ranatra sp.		×			×				Bu1,G2, Mi1	C
Family Notonectidae										
Buenoa arizonis									G2	C
Notonecta indica									G2	C
N. lobata									G2	C
N. undulata									G2	C
Order Megaloptera										
Family Corydalidae										
Corydalus cornutus				×		×	×	×	G2	C
Order Coleoptera										
Suborder Adephaga										
Family Dytiscidae										
Agabus seriatus	×		×		×				G2,G3	C
Bidessus sp.	×		×						G2,G3, Mi1	C
Deronectes aequinoctialis					×				G2,G3	C
D. nebulosus	×				×				G2,G3	C
D. roffi					×				G2	C
D. striatellus			×		×				G2,G3	C
D. yaquii					×				G2,G3	C
Dytiscus marginicollis					×				G2	C
Hydrophorus spp.	×	×			×				G2,Mi1	C
Laccophilus maculosus shermani									G2	C
L. pictus coccinelloides			×						G2,G3	C
Oreodytes sp.					×				G2	C
Rhantus sp.					×				G2	C
Thermonectus marmoratus		×			×				G2	C
Family Gyrinidae										
Dineutus sp.			×	×					G2	C

(continues)

TABLE 3.15   *Continued*

	Ms	Sc	T	D	V	S	G	C	Source[b]	Feeding Habits[c]
					Habitat[e]					
Gyrinus spp.			×	×	×	×			G2,Mi1	C
Family Haliplidae										
Haliplus sp.				×					G2	H
Peltodytes spp.	×	×	×	×					Bu1,G2, Mi1	CH
Suborder Myxophaga										
Family Hydroscaphidae										
Hydroscapha natans									G2	H
Suborder Polyphaga										
Family Dryopidae										
Helichus immsi						×	×	×	G2,G3	O
Family Elmidae										
Microcylloepus sp.	×			×			×	×	G2	O
Family Hydrophilidae										
Berosus exilis				×					G2	HO
B. punctatissimus	×	×		×					G2,G3	HO
Enochrus carinatus fucatus	×					×			G2	HO
E. picea glabrus						×			G2	HO
Hydrobius sp.	×								Bu1,G2	HO
Hydrochara lineata	×								G2	HO
Laccobius spp.	×	×		×					G2,Mi1	H
Tropisternus ellipticus	×	×							G2,G3	HO
Family Psephenidae										
Psephenus murvoshi								×	G2	O
Order Trichoptera										
Family Glossosomatidae										
Culoptila cantha	×						×	×	G2	O
Family Helicopsyche										
Helicopsyche borealis									G2	O
H. mexicanus	×					×	×	×	G2	O
Family Hydropsychidae										
Cheumatopsyche arizonensis	×						×	×	G2	O
Hydropsyche venada	×						×	×	G2	O
Family Hydroptilidae										
Ochrotricha spp.	×	×				×	×		Bu1,G2, Mi1	HO
Family Limnephilidae										
Limnephilus sp.		×				×	×	×	G2	O
Family Philopotamidae										
Chimarra augustipennis	×							×	G2	O
Family Polycentropodidae										
Polycentropus halidus	×	×	×			×	×	×	G2	O
Family Odontoceridae										
Marilia sp.		×							G2	O
Order Lepidoptera										
Family Pyralidae										
Paragyractis confusialis	×				×				G2	H
Order Diptera										
Suborder Nematocera										
Family Ceratopogonidae										

TABLE 3.15 *Continued*

	Ms	Sc	T	D	V	S	G	C	Source[b]	Feeding Habits[c]
				Habitat[e]						
Dasyhelea sp.	×					×			G2	O
Probezzia sp.	×	×			×	×			G2,G3, Mi1	C
Family Chironomidae										
Brillia sp.	×			×	×	×	×	×	G2	OD
Cardiocladius sp.	×					×			G2	C
Chironomus sp.					×	×	×		G2	OH
Corynocera sp.					×				G2,G3	O
Cricotopus sp.	×					×			G2,G3	O
Dicrotendipes sp.						×	×		G2,G3	O
Eukiefferiella sp.	×						×	×	G2,G3	OC
Micropsectra sp.	×					×			G2	O
Microtendipes sp.	×					×			G2	O
Paratendipes sp.						×			G2	O
Polypedilum sp.				×	×				G2	O
Tribelos sp.	×					×			F2,G2	O
unident. *Pentaneurini*	×	×	×			×			G2,Mi1	
Family Culicidae										
Culiseta incidens									G2	O
Family Simuliidae										
Simulium arcticum	×								G2	O
S. canadense	×								G2	O
S. virgatum	×								G2	O
Family Tipulidae										
Cryptolabis sp.	×					×			F2,G2, G3	OC
Tipula sp.				×					G2	O
Suborder Brachycera										
Family Dolichopodidae										
unident. spp.	×								G2,Mi1	
Family Ephydridae										
unident. spp.	×	×			×				Bu1,G2, Mi1	
Family Muscidae										
unident. spp.	×	×	×						Bu1,G2, Mi1	
Family Stratiomyidae										
Euparyphus sp.	×						×	×	G2,G3	O
Odontomyia (=Eulalia) sp.					×				G2	O
Family Syrphidae										
unident. spp.				×					G2	
Family Tabanidae										
Tabanus dorsifer	×					×			G2,G3	C
Phylum Chordata										
Subphylum Vertebrata										
Class Osteichthyes										
Order Cypriniformes										
Family Catosomidae										
Catostomus insignis									Ba,Me, Mi1, Mi2, S	O

(continues)

TABLE 3.15   *Continued*

	Habitat[e]							Source[b]	Feeding Habits[c]
	Ms	Sc	T	D	V	S	G	C		

	Ms	Sc	T	D	V	S	G	C	Source[b]	Feeding Habits[c]
Pantosteus clarki									Ba,Fl, Me, Mi1, Mi2, S	O
Family Cyprinidae										
Agosia chrysogaster									Ba,Fl, Me, Mi1, Mi2, S	O
Gila robusta									Ba,Me, Mi1, Mi2, S	O
Notropis lutrensis				×	×				Fl,Mi2	CO
Pimephales promelas				×					Fl,Mi2	O
Order Cypridontiformes										
Family Poeciliidae										
Gambusia affinis						×			Mi1,Mi2	O
Order Perciformes										
Family Centrarchidae										
Lepomis cyanellus						×			Mi1,Mi2	C

[a]An asterick indicates an introduced species.

[b]Autotroph habitats: Vg, vegetation (epiphytic); Fm, floating mat; Sl, slime; Sb, sediment benthos (epipelic); Hs, hard substrate (epilithic).

[c]Ba, Barber and Minckley (1966); Br, Bruns and Minckley (1980); Bu1, Busch (1979); Bu2, Busch and Fisher (1981); D, Deacon and Minckley (1974); Fl, Fisher et al. (1981); F2, Fisher and Gray (1983); F3, Fisher et al. (1982); G1, Gray (1980); G2, Gray (1981); G3, Gray and Fisher (1981); Me, Meffe and Minckley (1987); Mi1, Minckley (1981); Mi2, Minckley and Meffe (1987); S, Schreiber and Minckley (1981).

[d]Feeding habits: C = carnivore; CH = carnivore–herbivore; CO = carnivore–omnivore; D = detritivore; DO = detritivore–omnivore; H = herbivore; HO = herbivore–omnivore; O = omnivore; OC = omnivore–carnivore; OD = omnivore–detritivore.

[e]Invertebrate habitats: Ms, main stem; Sc, side canyon; T, tinajas; D, detritus; V, vegetation; S, sand; G, gravel; C, cobble (and boulder).

TABLE 3.16. *Structure of Functioning Ecosystem: Sycamore Creek, Sandy Benthos, Sand and Gravel Habitats*

Carnivores

Argia sp.
Cardiocladius sp.
Corydalus cornutus
Pantala hymenea
Physa virgata
Probezzia sp.
Progomphus borealis
Tabanus dorsifer

Omnivore–carnivores

Eukiefferiella sp.

Omnivores

Cheumatopsyche arizonensis	*Helicopsyche mexicanus*	*Oligochaete* spp.
Cricotopus sp.	*Hydropsyche venada*	*Paratendipes* sp.
Culoptila cantha	*Leptohyphes packeri*	*Pentaneurini* sp.
Dasyhelea sp.	*Limnephilus* sp.	*Polycentropus halidus*
Dicrotendipes sp.	*Microcylloepus* sp.	*Tribelos* sp.
Euparyphus sp.	*Micropsectra* sp.	*Tricorythodes dimorphus*
Helichus immsi	*Microtendipes* sp.	

(continues)

TABLE 3.16. *Continued*

Detritivores		Herbivores		
Detritivores	**Detritivore–omnivores**	**Herbivores**	**Herbivore–omnivores**	**Herbivore–carnivores**
	Baetis quilleri		*Chironomus* sp.	
	Brillia sp.		*Pantosteus clarki*	
	Thraulodes speciosus			

Detritus			Algae	
		Achnanthes exigua	*Gomphonema angustatum*	*N. secreta apiculata*
		A. lanceolata	*G. olivaceum*	*N. viridula avenacea*
		A. minutissima	*G. parvulum*	*Nitzschia amphibia*
		Amphora submontana	*G. tenellum*	*N. dissipata*
		A. venata	*Melosira varians*	*N. fonticola*
		Anabaena variabilis	*Meridion circulare*	*N. frustulum*
		Caloneis bacillum	*Navicula atomus*	*N. linearis*
		Cocconeis placentula var.	*N. cryptocephala*	*N. palea*
		lineata	*N. exiqua capitata*	*Nitzschia* spp.
		Cyclotella meneghiniana	*N. minima*	*Nostoc* sp.
		Cymbella hustedtii	*N. minuscula*	*Pinnularia viridus*
		C. minuta	*N. mutica*	*Surirella angustata*
		Epithemia sorex	*N. pupula*	*Synedra ulna*
		Fragilaria vaucheriae	*N. radiosa tenella*	

The sand and gravel was the habitat of several invertebrates: *Physa virgata*, a carnivore, and unidentified species of *Oligochaete*, which are probably omnivores. Unidentified species of arachnids were also present. Several insects were found associated with the sand and gravel. They were the carnivorous *Argia* sp.; *Progomphus borealis*, a carnivore; and *Pantala hymenea*, a carnivore. The mayflies present in this habitat were detritivore–omnivores, *Baetis quilleri* and *Thraulodes speciosus*; *Leptohyphes packeri*, an omnivore; and *Tricorythodes dimorphus*, also an omnivore. The sandy gravel was also a favorite habitat of the megalopteran *Corydalus cornutus*, a carnivore; the Dryopidan, *Helichus immsi*, an omnivore; and an unidentified species of the genus *Microcylloepus*, which is probably an omnivore. A few trichopterans were found in this habitat. They were the omnivorous *Culoptila cantha*, *Helicopsyche mexicanus* (a caddisfly), and *Cheumatopsyche arizonensis*, which is an omnivorous caddisfly, as is *Hydropsyche venada*. An unidentified species of *Limnephilus*, which is probably omnivorous, was found here, as was the omnivorous *Polycentropus halidus*. Several dipterans were in the sandy mud: the omnivorous *Dasyhelea* sp. and the carnivorous *Probezzia* sp. Other species in this habitat were the omnivore–detritivore *Brillia* sp., the carnivorous *Cardiocladius* sp.; *Chironomus* sp., probably an omnivore–herbivore; and the omnivorous *Cricotopus* sp. and *Dicrotendipes* sp. An omnivore–carnivore, *Eukiefferiella* sp., was also present, as were the omnivorous *Micropsectra* sp., *Microtendipes* sp., and *Paratendipes* sp. The latter three are, of course, chironomids or tendipedids. Another omnivore species present is *Tribelos* sp. An unidentified species of the genus *Pentaneurini* was also present. The dipteran *Euparyphus* sp., an omnivore, was also found among the sand and gravel, as was *Tabanus dorsifer*, a carnivore.

Several fish were found in these various habitats. In Sycamore Creek, *Pantosteus clarki*, a herbivore with omnivorous tendencies, was common. (Fisher et al., 1981).

Lotic Habitats: Rubble, Rocks, and Pebbles. The hard substrates, such as rubble, rocks, and pebbles, often found in fast-flowing water, were habitats for a number of species. Attached to these hard substrates were blue-green algae, *Oscillatoria lutea*, *Schizothrix calcicola*, and *Calothrix parietina*. The filamentous green algae, so common in the Grand Canyon, was found attached to rocks in these Sonoran Desert streams: *Cladophora glomerata*. Also present was *Zygnema* sp. Several diatoms were found attached to these hard surfaces: *Cyclotella meneghiniana*, *Melosira varians*, *Meridion circulare*, and *Synedra ulna*. Closely attached in fairly fast flowing water were *Achnanthes lanceolata*, *A. minutissima*, and *Cocconeis placentula* var. *lineata*. Other diatoms attached to these hard surfaces were *Amphora submontana*, *A. venata*, *Cymbella hustedtii*, and *C. minuta*. Also attached, probably where the current was less strong, were *Gomphonema angustatum*, *G. lanceolatum*, *G. olivaceum*, *G. parvulum*, and *G. tenellum*. Other diatoms found associated with these hard substrates, probably in between the pebbles and rocks, were *Navicula cryptocephala*, *N. exigua capitata*, *N. minima*, *N. minuscula*, *N. mutica*, *N. pupula*, *N. radiosa tenella*, *N. secreta apiculata*, and *N. viridula avenacea*. Belonging to the family Nitzschiaceae was *Hantzschia amphioxys*. Other taxons belonging to this family found on these hard substrates were five taxons belonging to the genus *Nitzschia* (Table 3.15 and 3.17). Belonging to

TABLE 3.17. *Structure of Functioning Ecosystem: Sycamore Creek, Lotic Habitats—Rubble, Rocks, and Pebbles*

―――――――――――Carnivores―――――――――――

Argia sp.
Corydalus cornutus
Lepomis cyanellus
Physa virgata
Tabanus dorsifer

―――――――――Omnivore–carnivores―――――――――

Eukiefferiella sp.
Notropis lutrensis

―――――――――――Omnivores―――――――――――

Cheumatopsyche arizonensis   *Helicopsyche mexicanus*
Chimarra augustipennis   *Hydropsyche venada*
Culoptila cantha   *Limnephilus* sp.
Dicrotendipes sp.   *Microcylloepus* sp.
Euparyphus sp.   *Pimephales promelas*
Gambusia affinis   *Psephenus murvoshi*
Helichus immsi

Detritivores		Herbivores		
Detritivores	**Detritivore–omnivores**	**Herbivores**	**Herbivore–omnivores**	**Herbivore–carnivores**
Mesocapnia arizonensis	*Brillia* sp.		*Chironomus* sp.	
	Thraulodes speciosus			
	——Detritus——		——Algae——	
		Achnanthes lanceolata	*G. olivaceum*	*N. secreta apiculata*
		A. minutissima	*G. parvulum*	*N. viridula avenacea*
		Amphora submontana	*G. tenellum*	*Nitzschia dissipata*
		A. veneta	*Hantzschia amphioxys*	*N. fonticola*
		Calothrix parietina	*Melosira varians*	*N. frustulum*
		Cladophora glomerata	*Meridion circulare*	*N. linearis*
		Cocconeis placentula var. *lineata*	*Navicula cryptocephala*	*N. palea*
		Cyclotella meneghiniana	*N. exiqua capitata*	*Oscillatoria lutea*
		Cymbella hustedtii	*N. minima*	*Rhopalodia gibba*
		C. minuta	*N. minuscula*	*Schizothrix calcicola*
		Gomphonema anqustatum	*N. mutica*	*Surirella anqustata*
		G. lanceolatum	*N. pupula*	*Surirella ovata*
			N. radiosa tenella	*Synedra ulna*
				Zygnema sp.

the family Epithemiaceae was *Rhopalodia gibba*. The order Surirellales was represented by two taxons belonging to the genus *Surirella* in this habitat.

Associated with these habitats, consisting mainly of gravel, rubble, and rocks, were several animals: a snail, *Physa virgata*, which is a carnivore; the odonate *Argia* sp., also a carnivore; a mayfly, *Thraulodes speciosus*, which is a detritivore–omnivore; and a stonefly, *Mesocapnia arizonensis*, a detritivore; the megalopteran *Corydalus cornutus*, a carnivore; beetle *Helichus immsi*, an omnivore; and an unidentified species of the genus *Microcylloepus* sp., an omnivore. Another Coleoptera in this habitat was the omnivorous *Psephenus murvoshi*. The trichopterans were *Culoptila cantha*, an omnivore; *Helicopsyche mexicanus*, an omnivore; *Cheumatopsyche arizonensis*, an omnivore; and *Hydropsyche venada*, an omnivore. Other species belonging to this order were *Limnephilus* sp., an omnivore; *Chimarra augustipennis*, an omnivore; and the dipterans *Brillia* sp., an omnivore–detritivore; *Chironomus* sp., an omnivore–herbivore; *Dicrotendipes* sp., an omnivore; and *Eukiefferiella* sp., an omnivore–carnivore. Other insects in this habitat were the dipteran *Euparyphus* sp., an omnivore, and *Tabanus dorsifer*, a carnivore.

Several species of fish were found among these rocks in this rapid habitat: *Notropis lutrensis*, a carnivore–omnivore; *Pimephales promelas*, an omnivore; and in the sandy areas, *Gambusia affinis*, an omnivore, which was mainly in backwaters, as was *Lepomis cyanellus*, a carnivore.

SUMMARY

The Gila River system streams are typical of arid area streams. They are characterized by large watershed areas that amplify the limited precipitation inputs, which form destructive flash floods. Yearly intermittent surface flows, high evaporation, and substantial subsurface flows are other physical characteristics.

The water of these streams is hard, alkaline, and somewhat saline; because of sufficient nutrient levels and high light levels, these streams support extensive algae production and are largely autotrophic. The biota is greatly reduced by the destructive floods, but they recover rapidly. These streams differ from most headwater streams in various parts of the United States in that they are classed as autotrophic rather than heterotrophic.

Two well-studied and relatively undisturbed streams are Aravaipa Creek, a tributary of San Piedro River, and Sycamore Creek, a tributary of the Verde River.

Physical Characteristics.   The Aravaipa Creek is an unregulated middle- to low-elevation perennial stream with a watershed area of 1400 km^2. It can be divided into three physiographic sections: the upper section, the middle section, and the floodplain section. The upper part of the Aravaipa watershed is vegetated by desert grasslands, while the lower creek flows through Sonoran Desert scrub. Mature riparian forest is found in the upper valley. This community grades into one dominated by willows, cottonwood, and seep willow.

Pools typically occur next to cliffs and adjacent obstructions. The bottom substrate of the stream consists of riffles and runs separated by various graded channels flowing over sand and gravel. Rapids are relatively common in the canyon. Outside the canyon, the channel is wider, shallower, and often braided. Pools are shallower but more common and rapids uncommon. The sediment consists of unconsolidated and poorly sorted sand, gravel, pebbles, cobbles, and boulders.

Turbidity usually has bimodal peaks in winter and in summer. Winter storms are more severe than summer. Discharge also has bimodal peaks. Floods originating in the upper valley carry large volumes of silt, sand, and gravel. When discharge is below 1 m^3 s^{-1} for at least 10 days, the turbidity is less than 1.0 JTU. Winter floods seem to carry more semicolloids and colloids than do summer floods.

The temperature variations are quite large. From June to August it is about 9.6°C, while the December through February mean is about 6.7°C. Summer temperatures are coolest in heavily shaded canyons. In the upper valley the water is cooler in summer and warmer in other seasons than below the canyons. Presumably because of greater subsurface flow and groundwater inflows in the upper valley, the mean January stream temperature in the upper Aravaipa is 12.5°C. In the lower Aravaipa the mean temperature is 8°C.

Sycamore Creek has a watershed of 505 km^2 with altitudes ranging from 427 to 4164 m above sea level. As with the Aravaipa Creek, the terrain is mountainous and rugged. Summer modal discharge is 0.03 m^3 s^{-1}, and the mean modal current velocity is 0.015 m/s. Sand and gravel are the predominant substrates.

Chemical Characteristics. The Aravaipa Creek is hard to very hard and has relatively high concentrations of sodium. Probably as a result of photosynthetic bicarbonate scavenging, calcium concentrations decreased slightly downstream, whereas sodium increases downstream. Mercury concentrations averaged 5.34 μg L^{-1} and ranged from 0 to 75 μg L^{-1}. It sometimes exceeds the U.S. Environmental Protection Agency water quality standard, which is 0.05 μg L^{-1}. Cadmium concentrations range from 1.3 to 1.34 μg L^{-1}. The main-stream cadmium concentrations averaged 24.72 μg L^{-1} and ranged from 1.3 to 134 μg L^{-1}, exceeding the water quality standards. pH varies from 7.4 to 8.9. pH is lower in the summer and at night. Nitrogen and orthophosphate-P concentrations are generally higher after floods and during winter. Downstream decline in NO_3-N in desert streams is largely due to autotrophic uptake. The upper Aravaipa daytime dissolved oxygen is highest in late winter and lowest in late spring and summer.

Sycamore Creek has moderately hard calcium–magnesium bicarbonate waters. The summer pH varies from 7 to 8. N as nitrate usually exceeds 0.10 mg L^{-1}, and dissolved phosphorus is generally below 0.050 mg L^{-1}. Dissolved oxygen in interstitial water of Sycamore Creek is generally slightly lower than surface water. Anoxic conditions were never found even at depths.

Primary Production. Standing crops of detritus are greatest after flooding. The mean values for the Aravaipa Creek were 25.8 to 47.0 g m^{-2} immediately after floods and 1.7 to 7.1 g m^{-2} with 45 or more days of modal discharge. Winter and spring levels of detritus are relatively low, whereas summer levels are consistently high and autumn levels reflect the residual effect of the summer peaks. Floods decrease the amount of detritus. In Sycamore Creek the standing crop of detritus was 231.6 g m^{-2} before summer flooding and 75.7 g m^{-2} for 16 to 45 days after flooding.

In Sycamore Creek the stream detritus is primarily FPOM; and 82 to 94% of all POM was smaller than 100 μm. The bulk of flood detritus is FPOM of terrestrial origin. Riparian leaf fall usually occurs in December, but no significant accumulations have been observed in the stream. This is because much of the autumnal leaf fall in desert streams impinges on a band of dry creek bed separating the stream from the riparian vegetation. Of the accumulated leaf fall that does impinge in the stream, much is probably exported because desert stream channels typically have few retention devices. In a few days of a flood, the detritus standing crop declines via downstream or lateral export or in situ assimilation. During this decline diatoms and filamentous algae colonize the newly scoured substrates, and as they increase the detritus derived from these autochthonous producers becomes the major source of FPOM. Primary production is mainly from algae, which is relatively high in Aravaipa Creek during periods of modal discharge often exceeding those recorded for temperate streams.

The mean biomass of epipelic algae in Aravaipa Creek was 0.21, 0.01, and 2.0 g m^{-2} of ash-free dry weight for the upper valley, canyon, and lower valley. For epilithic assemblages the means were 1.48, 0.41, and 1.62 g m^{-2}. For the *Cladophora* epiphytic assemblages the means were 15.3, 3.2, and 26.4 g m^{-2}, respectively. Shading accounted for the lower amount of *Cladophora* and other algae in the canyon. Algae biomass in Aravaipa Creek was greatly reduced immediately after flooding. Epilithic algae never exceeded 0.5 g m^{-2} ash-free dry weight and the *Cladophora* assemblage never exceeded 6.0 g m^{-2} within 15 days of a spate. This reduction of algal biomass reflects the destructiveness of summer flooding.

Algae biomass in Sycamore Creek is also high. Summer preflood epipelic patches dominated by diatoms had mean patch ash-free dry weights of 174 mg m^{-2} and mean chlorophyll *a* concentrations of 59 mg m^{-2}. Patches of *Cladophora* plus diatom epiphytes had mean ash-free dry weight of 363 mg m^{-2} and chlorophyll *a* concentrations of 324 mg m^{-2}. This was in the summer. In October, the ash-free dry weight and chlorophyll *a* concentrations recorded for the same patch type were 240 g m^{-2} and 143 mg m^{-2}, respectively. Ash-free dry weight and chlorophyll *a* concentrations for blue-green algae were 170 g m^{-2} and 140 mg m^{-2}, respectively. For *Cladophora* plus epiphytes plus blue-green algae patch they were 310 g m^{-2} and 190 mg m^{-2}.

The primary production in the Sonoran Desert streams is high. Primary production in Sycamore Creek was estimated to be 8.5 g O$_2$ m^{-2} per day. The community respiration (CR) was estimated to be 5.1 g O$_2$ m^{-2} per day and the PR ratio (GPP/CR) was 1.7 g O$_2$ m^{-2} per day. With the same regression model, Fisher et al. (1982) found that both GPP and CR increased asymptotically in Sycamore Creek after a summer flood. By 30 days postflood, GPP was 6.6 g O$_2$ m^{-2} per day. The CR (community respiration), about 22%, is attributed to insects. A large part of the rest is assignable to plants and probably only a small amount to fish. Regardless of the difficulties involved in estimating relationships between allochthonous and autochthonous inputs and exports, biomass does increase throughout the three-month postflood period, although the rate of increase appears to level off.

Secondary Production.   In Aravaipa Creek invertebrates were more numerous and have greater biomass in the upper valley, while the lowest mean density and biomass occurred in the canyon. Density usually ranged from 1×10^5 to 2×10^5 individuals m^{-2}, and biomass generally ranged from 10 to 20 g m^{-2} wet weight. The invertebrate abundance is strongly associated with filamentous algae detritus and a coarse sand substrate that supports a well-diversified interstitial fauna. Flooding dramatically affects invertebrate populations. *Baetis* sp., *Callibaetis* sp., *Tricorythodes* sp. (Ephemeroptera), Chironomidae, and Simuliidae are abundant year round. Oligochaeta, Acarina, and Ostracoda were numerous, and Hydropsychidae were major contributors to the biomass. *Ephemerella micheneri* was abundant during the spring, and *Masocapnia frisoni*, a stonefly, was abundant during the winter.

In Aravaipa Creek, many short-lived invertebrates such as Ephemeroptera and Chironomidae usually recover rapidly from floods. But extended periods of rainfall and flooding (e.g., the winter of 1978) can inhibit their recovery. Sustained flooding from July 1977 through April 1978 in Aravaipa Creek virtually eliminated the longer-lived insects such as the Trichoptera and greatly reduced populations of Odonata, Hemiptera, and Coleoptera. *Cladophora* mats are important sources of food and habitat for invertebrates.

In Sycamore Creek, the mean invertebrate biomass is 3 g m^{-2} with a range from 0 to 10 g m^{-2}. The secondary production of the collector component of macroinvertebrate communities in Sycamore Creek is high compared with other streams. Secondary production in Sycamore Creek is dominated by a few taxa from the Ephemeroptera and Diptera and Chironomidae. Thus species-specific annual production rates are also high. As an example, annual production of *Baetis quilleri* is about 10-fold higher (21.9 g m^{-2} per year) than other species of *Baetis* in cooler streams (2.1 g m^{-2} per year). Sycamore Creek flooding in the summer had a severe effect on the invertebrate biomass. It took 13 days for the biomass to reach 50% of preflood levels.

Trichoptera were extirpated from Sycamore Creek by extensive flooding from November to April. It is evident that collectors ingest just about 4.2 times their body mass per day. Considering that the summer gross primary production is 8.5 g m^{-2} per day, Sycamore Creek collectors ingest about 1.5 times the primary production during a two-month postflood recovery period. Collectors ingested organic matter equivalent to 2.5 to 6.2 times the postflood primary production. Sycamore Creek invertebrates are estimated to turn over total organic matter every two to three days, much faster than the 71 to 102 days estimated for an English chalk stream. Since primary production by the dominant, *Cladophora glomerata*, is resistant to processing until it becomes detritus, other fractions of the organic matter pool must be processed at an even higher rate. The extensive organic matter processing by Sycamore Creek invertebrates occurs because of a high ingestion rate (44 to 251 μg mg^{-1} per hour) and exceedingly low assimilation efficiency 7 to 15%). Rapid growth of Sycamore invertebrates combined with high ingestion rates and low assimilation efficiency is comparable to that in temporary habitats. The strategy may be to shorten gut retention time, absorb only the most labile fraction, egest fecal material for microbial conditioning, and reingest this microbial fecal complex.

In Aravaipa Creek the mean fish density was highest in the upper valley, lowest in the canyon, and intermediate below the canyon. The canyon and upper valley had similar annual mean fish biomass (21.0 g m^{-2}), but their seasonal patterns were different. The lower valley had a mean of 14.4 g m^{-2}. There were seasonal peaks during the winter above and below the canyon, while the canyon peaked in the summer. The upper valley was low during the summer, while the canyon was low during spring and autumn. Flooding has a less direct effect on fish populations than it does on algae and invertebrates. This varied according to the species of fish.

   Communities of Aquatic Life. Desert streams do not have large accumulations of allochthonous CPOM. Consequently, invertebrate detritivores, shredder fauna, is represented primarily by only two genera: *Brillia* sp. during the summer and *Mesocapnia* spp., a plecopteran, in the winter. FPOM detritivores and detritivore–herbivores dominated the invertebrate fauna of both Aravaipa and Sycamore Creeks. Collector–gatherers accounted for 87% of the density and 85% of the biomass in Sycamore Creek. Collector–gatherers and scrapers contributed about 76% to the total population densities in Aravaipa Creek.

   Collector–Gatherer Insects. Allochthonous detritus is a collector's diet for the first 10 days after the flood. As diatom biomass increases, diatoms become the dominant diet. Several weeks later, autochthonous detritus becomes about as important as diatoms to their diet. Filamentous algae is not an important diet component. It is known that the dominant filamentous algae, *Cladophora glomerata*, contains lauric acid, which is believed to be harmful to many invertebrates. Omnivore collector filterers are second in terms of density at Aravaipa Creek, contributing about 12% of the fauna. Invertebrate carnivores accounted for about 13% of the density and 10% of the biomass during postflood periods in Sycamore Creek. Carnivores accounted for 77% of the total density of invertebrates in Aravaipa Creek. Chironomidae, ceraptogonids, and distichids were the common carnivores.

   Among the fish in Aravaipa Creek are *Pantosteus clarki*, a herbivore, *Agosia chrysogaster*, an omnivore, and the remaining fish are carnivores. During the summer, there is considerable dietary overlap between the carnivores and the omnivores, which are mainly fish. Ephemeroptera, predominant in the benthos and drift, are the most heavily utilized prey items. Of the mayflies, Baetidae are preyed upon most heavily. The difference in predation on other Ephemeroptera reflects the carnivorous habit preference and prey seasonality. Seasonal peaks of Plecoptera and Trichoptera are also reflected in fish diets. Preferential feeding of Diptera, mainly Chironomidae and Simuliidae, is evident for the bottom-feeding fish, although other fish also feed on Diptera. There is a decrease in volume of food in fish stomachs during the summer. It is not clear whether this is due to decreased consumption or to the increased gut evacuation rates usually associated with high temperatures.

   In Sycamore Creek two species of fish, *Pantosteus clarki* and *Agosia chrysogaster*, are herbivores. When food is plentiful, spatial and temporal food resource partitioning is evident for autumn populations of these two species. *A. chrysogaster* fed on and around mats of a filamentous alga, *Cladophora glomerata*. *Zygnema*, epiphytic diatoms, and some epipelic diatoms were common in the gut. *P. clarki* probably fed in places where filamentous algae were less common. Epipelic diatoms are the common gut content. *A. chrysogaster* fed during the day, while *P. clarki* had feeding peaks at dusk and dawn. When food was scarce, food partitioning between these two species was not as evident.

Primary Producers. Of primary producers in both Aravaipa and Sycamore Creeks, *Cladophora glomerata* was the most abundant filamentous algae. Epiphytic diatoms on *Cladophora* are also a very important source of food. Diatoms are a very important source of food to both vertebrates and invertebrates in these creeks. *Cladophora glomerata* is also important. Over time the algal patches change. The epipelic patches of algae are relatively persistent but have a 27% chance of being overgrown by blue-green algae. Blue-green patches have a similar chance, 33%, of persistence, changing to a diatom patch, and they, in turn, of developing into a *Cladophora*, blue-green, and epiphytic diatom patch.

Flooding. Few invertebrates are resistant to flooding, but many are resilient; that is, they recover quickly. Certain Hemiptera and adult Coleoptera are resilient. Species of *Baetis*, *Chloroterpes inornata*, Chironomidae, and *Oligochaeta* are resilient, whereas *Tricorythodes* sp. (Ephemeroptera) and *Ochrotrichia* (Trichoptera) were not. Coleoptera, Hemiptera, and Odonata became very uncommon in Aravaipa Creek after flooding. Trichoptera were not collected after flooding in Sycamore Creek. Recolonization was probably from resources outside the watershed. Populations that seemed to recover more quickly were *Baetis* spp., Chironomidae, and Simuliidae. Numerically important taxa that were vulnerable to summer flooding were *Chloroterpes* sp., *Tricorythodes* spp. (Ephemeroptera), Glossosomatidae (Trichoptera), Ceratopogonidae (Diptera), *Acarina*, Ostracoda, and Planaridae. Taxa that were vulnerable to both winter and summer flooding include the Hydropsychidae, the Hydroptilidae, and the Oligochaeta. Taxa that are vulnerable, most particularly the Trichoptera, became abundant in the absence of flooding. Those that were vulnerable to extensive winter flooding were *Ephemerella micheneri* and *Mesocapnia frisoni*, and a species seasonably restricted was *E. micheneri*. A species vulnerable to late winter and spring flooding was *Mesocapnia frisoni*, which was resistant to winter flooding.

In Sycamore Creek most of the Ephemeroptera, Diptera, and Hemiptera have development times and degree-day requirements that are among the lowest recorded for their respective orders. Reproduction is continuous, and the number of generations per year for some of these taxa may exceed 35. These short development times probably allow the aerial adults to recolonize after flooding. The Trichoptera reproduce year-round but are usually more successful in autumn and spring, when flooding is unlikely. In Sycamore Creek, the trichopterans have developed shorter generation times than is characteristic of comparable taxa from temperate streams. However, they have similar day requirements. Unlike most insects, the Coleoptera exhibit strong seasonality in Sycamore Creek. Concentrating their larvae development in spring and early summer or in late summer and early autumn, they usually avoid the worst effects of flooding.

Longer-lived taxa usually avoid drought by selectively ovipositing in persistent aquatic habitats. Adult aerial dispersal is a strategy employed by most stream insects as a primary recolonization pathway. Behavioral drift, particularly among immature Ephemeroptera and Chironomidae, enhances recolonization by dispersing insects from localized and aggregated ovipositional sites. Drift provides places to recolonize. Various habitats in the Sycamore Creek support different associations of insects.

Several species of native fish are present in Aravaipa Creek. Introduced species are not well established. Only the green sunfish is able to build populations, usually in the lower reaches of the creek. In Aravaipa Creek seven native species have been persistent. They have behavioral and morphological adaptations that help them to maintain their position during flooding. On the other hand, introduced species have evolved under lowland mesic conditions, where downstream displacement or lateral escape onto floodplains are feasible responses to flooding.

Functional Ecosystems: Aravaipa Creek.   Functional ecosystems of aquatic life are found in the vegetative habitats, sandy benthos, and sandy gravel habitats and in lotic habitats (i.e., hard substrate habitats).

Vegetative Habitats.   In Aravaipa Creek, the vegetative habitats are found among floating or attached plants and slime, which was algae—probably blue-greens— growing attached to sediment particles in the bottom of pools or slow-flowing water. Those organisms found attached to plants were green algae and diatoms. Crawling over the plants were *Physa* and unidentified species of Oligochaetes, which are probably omnivores. Other animals were Crustacea, omnivores, and an unidentified species belonging to the genus *Cypris*. Among the vegetation were Odonates, which are probably carnivores. There were also a few mayflies, which are probably detritivore–omnivores, and the omnivorous mayfly *Callibaetis* sp. Other species in this habitat were *Mesocapnia frisoni*, which is a detritivore–omnivore, and the hemipteran *Abedus herberti*, a carnivore. Also crawling over the plants were beetles, both omnivores and carnivores. Several hydrophilids were found associated with the vegetation. They were herbivore–omnivores, and *Berosus* sp. and *Laccobius* sp. were herbivores. Several trichopterans were found which are probably herbivore– omnivores. A dipteran, *Probezzia* sp., which is a carnivore, was found in this habitat.

Sandy Benthos Habitat.   In sand and gravel habitats in the benthos were found green algae and diatoms. Found in this habitat were the omnivorous *Euglena* sp. and *Phacus* sp.; a flatworm that is probably a carnivore; and a species belonging to the phylum *Nemata* and also an unidentified species of the genus *Physa*, which are probably omnivores, although some tend to be carnivorous. Oligochaetes that are probably omnivores were found in this habitat, as were ostracods, which are probably omnivores. Also found are several species of odonates, which are carnivorous, and of mayflies of the genus *Baetis*, which are detritivore–omnivores; *Heptagenia* sp., an omnivore; *Rhithrogena* sp., an omnivore–detritivore; *Choroterpes* sp., an omnivore; a species of *Ephemerella* that is probably an omnivore; and Hemiptera that were carnivores. Also present were beetles that are carnivores. Omnivores were trichopterans belonging to the genera *Cheumatopsyche* and *Hydropsyche*; a species belonging to the genus *Ochrotricha*, which is an herbivore–omnivore; and the omnivore–carnivore species *Rhyacophila*. Carnivorous Diptera were also present, as were unidentified species of chironomids belonging to the genus *Tanytarsini*, which are omnivores. Species belonging to the genus *Pentaneurini*, which are carnivores–omnivores, were also present as were blackfly larvae that tend to be omnivores.

Lotic Habitats. In lotic habitats on hard substrates were blue-green algae, *Oscillatoria* sp. and *Gleocapsa* sp. Green algae present were unidentified species of *Tetraspora, Microspora,* and *Gongrosira.* Other green algae found on these hard substrates were species belonging to the genera *Ulothrix, Stigeoclonium, Cladophora, Rhizoclonium,* and *Zygnema.* Diatoms were also present, as were a few specimens of *Chara* sp.

Crawling over the rocks were a few species of mayflies that were detritivore–omnivores or omnivores; a megalopteran, *Corydalus* sp., a carnivore; and beetles that were omnivores, as well as hydrophilids, trichopterans, *Cheumatopsyche,* and *Hydropsyche.* Omnivore–carnivores included *Rhyacophila* sp., a few chironomids were present that are probably omnivores, and blackfly larvae there are probably omnivores.

Functional Ecosystems: Sycamore Creek

Vegetative Habitats. In Sycamore Creek we found a similar structure in the food web or ecosystem. In vegetative habitats there were diatoms that were the main base of the food web. Invertebrates found were a species of *Physa,* a carnivore, sometimes an omnivore; odonates, which were carnivores; a mayfly, which is a carnivore–omnivore; another mayfly, *Caenis* sp., an omnivore; hemipterans which are carnivores; and a few beetles which were herbivores. Also present were corixids, which are carnivores. Several species of beetles were found in and among the vegetation which were carnivores. However, there were specimens of the herbivorous *Haliplus* sp. and the carnivore–herbivore genus *Peltodytes.* Other herbivore–omnivores belong to the genus *Berosus* and the herbivorous *Laccobius.* Other species belonging to the Trichoptera were herbivore–omnivores. Chironomids were present that were omnivore–detritivores and omnivore–herbivores and others that are strictly omnivorous.

Sand Habitats. In sand habitats were found blue-green algae and several species of diatoms. The invertebrates associated with this habitat were *Physa virgata,* unidentified species of Oligochaetes, and spiders. Insects found associated with the sand and gravel were several species of carnivorous odonates, mayflies which were detritivore–omnivores and omnivores; Megaloptera which were carnivores and omnivores; caddisflies which are probably omnivores; and Diptera which are omnivores, carnivores, omnivore–detritivores, and omnivore–carnivores. Several species of fish were found in this habitat.

Several fish were found in these various habitats. In Sycamore Creek, *Pantosteus clarki,* a herbivore with omnivorous tendencies, was common (Fisher et al., 1981).

Lotic Habits: Gravel, Rocks, and Pebbles. Attached to these hard rocks were various species of blue-green algae, particularly *Oscillatoria leutea, Schizothrix calcicola,* and *Calothrix parietina.* The common green algae was *Cladophora glomerata.* Also present was *Zygnema,* several species of diatoms were found in and among these algae. The main animals found in this habitat associated with gravel, rubble, and rocks were a snail, *Physa virgata;* odonates, mayflies, stoneflies, Megaloptera, beetles, and caddisflies, as well as several dipteran species. These species were detritivore–omnivores, detritivores, and omnivore–herbivores. There were also omnivores and omnivore–carnivores, as well as carnivores. They are all diversified and belong to the major groups of insects. Several species were found in among the rocks in this rapid habitat: *Notropis lutrensis*, a carnivore–omnivore; *Pimephales promelas*, an omnivore; and *Gambusia affinis*, an omnivore that was in the more sandy areas. In backwaters was the carnivore *Lepomis cyanellus*.

BIBLIOGRAPHY

Barber, W. E., and W. L. Minckley. 1966. Fishes of Aravaipa Creek, Graham and Pinal Counties, Arizona. Southwest. Nat. 11: 313–324.

Bruns, D. A. R., and W. L. Minckley. 1980. Distribution and abundance of benthic invertebrates in a Sonoran Desert stream. J. Arid Environ. 3: 117–131.

Busch, D. E. 1979. The patchiness of diatom distribution in a desert stream. J. Ariz.-Nev. Acad. Sci. 14: 43–46.

Busch, D. E., and S. G. Fisher. 1981. Metabolism of a desert stream. Freshw. Biol. 11(4): 301–307.

Deacon, J. E., and W. L. Minckley. 1974. Desert fishes. *In* G. W. Brown, ed., Desert biology, vol. 2. Academic Press, San Diego, Calif.

Fisher, S. G. 1986. Structure and dynamics of desert streams. pp. 119–139. *In* W. G. Witford, ed., Patterns and process in desert ecosystems. Univ. New Mexico Press, Albuquerque, N. Mex.

Fisher, S. G., and L. J. Gray. 1983. Secondary production and organic matter processing by collector macroinvertebrates in a desert stream. Ecology 64(5): 1217–1224.

Fisher, S. G., D. E. Busch, and N. B. Grimm. 1981. Diel feeding chronologies in two Sonoran Desert stream fishes, *Agosia chrysogaster* (Cyprinidae) and *Pantosteus clarki* (Catostomidae). Southwest. Nat. 26(1): 31–36

Fisher, S. C., L. J. Gray, N. B. Grimm, and D. E. Busch. 1982. Temporal succession in a desert stream ecosystem following flash flooding. Ecol. Monogr. 52(1): 93–110.

Gray, L. J. 1980. Recolonization pathways and community development of desert stream macroinvertebrates. Unpublished dissertation. Arizona State Univ.

Gray, L. J. 1981. Species composition and life histories of aquatic insects in a lowland desert stream. Am. Midl. Nat. 106(2): 229–242.

Gray, L. J., and S. G. Fisher. 1981. Postflood recolonization pathways of macroinvertebrates in a lowland desert stream. Am. Midl. Nat. 106(2): 249–257.

Grimm, N. B., and S. G. Fisher. 1984. Exchange between interstitial and surface water: implications for stream metabolism and nutrient cycling. Hydrobiologia 111(3): 219–228.

Meffe, G. K., and W. L. Minckley. 1987. Persistence and stability of fish and invertebrate assemblages in a repeatedly disturbed Sonoran Desert stream. Am. Midl. Nat. 117(1): 177–191.

Minckley, W. L. 1981. Ecological studies of Aravaipa Creek, central Arizona, relative to past, present and future uses. Final Report. U.S. Bureau of Land Management, Safford, Ariz. 362 pp.

Minckley, W. L., and G. K. Meffe. 1987. Differential selection for native fishes by flooding in streams of the arid American Southwest. pp. 93–104. *In* W. J. Matthews and D. C. Heins, eds., Ecology and Evolution of North American Stream Fish Communities. Univ. of Oklahoma Press, Norman, Okla.

Schreiber, D. C., and W. L. Minckley. 1981. Feeding interrelationships of native fishes in a Sonoran Desert stream. Great Basin Nat. 41: 409–426.

Sumner, W. P., and C. D. McIntire. 1982. Grazer–periphyton interaction in laboratory streams. Arch. Hydrobiol. 93: 135–157.

Vollenweider, R. A. 1974. Sampling techniques and methods for estimating quantity and quality of biomass. *In* R. A. Vollenweider, ed., A manual on methods for measuring primary productivity in aquatic environments. Blackwell Scientific, Oxford.

CHAPTER 4

Water Quality Problems

HARDNESS

Total hardness, a parameter closely related to salinity, is of primary interest in assessing water quality. Increase in concentration of hardness leads to added soap and detergent consumption, corrosion and scaling of metal water pipes and water heaters, and accelerated fabric wear. However, if you soften the water, costs increase. After a certain degree of increase in hardness, the water is unusable for these uses. Water from the Colorado River below Lake Mead would be classified as very hard. Boiler feed and coolant water comprise a major portion of water use to industry in the basin. The mineral quality of boiler feedwater is an important factor in the rate of scale formation on heated surfaces, degree of corrosion in the system, and quality of produced steam. In cooling-water systems, resistance to slime formation and corrosion are affected by mineral quality. The quality of water is maintained in boiler and cooling systems, by reducing the mineral content to relatively good quality water.

SALINITY

The salinity conditions of the Colorado River have always been severe. Naturally, it is a very saline river; over 9 million tons of salt is carried down the river every year. This salt is diluted by the flow of the river, and depending on the quantity of flow, salinity can be relatively dilute or concentrated. The main effect of this high salinity is economic. The U.S. Geological Survey monitors the flow and salinity of the river system continuously through a network of 20 gauging stations. As the impact of recent water development works its way through the hydrologic systems, salinity will increase. As development continues, more salinity control will be necessary. Salinity in the Colorado River basin is influenced directly by reservoir storage, water resource development, salinity control, climatic conditions, and natural runoff.

If the present use and diversion of water continues, there will have to be significant increases in salinity control. At present 621,400 tons of salinity must be controlled. By the year 2015 we will have to control 1,476,600 tons, more than twice the amount controlled by present measures. (U.S. Department of the Interior, 1997).

Natural Sources of Salinity
The natural sources of salinity are saline springs. These sources exist over long reaches of the riverine system. Salt pickup occurs over large surface areas from underlying soils, geologic formations, and stream channels and banks. It is believed that natural sources contribute the largest overall share of salts in the Colorado River (Figure 4.1). The natural salt load of the Colorado River at Lees Ferry, Arizona has been estimated to be about 5.3 million tons per year. Natural point sources are saline springs, where the contribution of salt and water is easily identified if it comes from a single source or from concentrated sources. Two examples are the Glenwood–Dotsero Springs Unit, which contributes 440,000 tons/yr, and the Paradox Valley Unit area, which contributes 205,000 tons/yr. These are but two examples of point sources that contribute large amounts of salt. It is evident that large natural salt loads occur in certain reaches of the Colorado River that can not be attributed to irrigation or other development-related activities. These are due primarily to natural diffuse sources and saline springs. The pickup of these large diffuse sources of salt occurs over large surface areas from underlying soils, geologic formations, and stream and channel banks.

Natural sources contribute about two-thirds of the average annual salt load passing Hoover Dam. However, relatively small areas such as Paradox Valley have very high rates of pickup and contribute large salt loads. Diffuse sources contribute about half of the basin's salt burden. Discrete or point salinity sources also occur throughout the basin. In the Lower Basin, mineral springs add more salt to the Colorado River than does any other type of salinity source. For example, Blue Springs, located near the mouth of the Little Colorado River, contributes a salt load of about 547,000 tons per year, approximately 5% of the annual salt burden at Hoover Dam. Blue Springs is the largest point source of salinity in the entire Colorado River basin. In the Upper Basin some 30 significant mineral springs have been identified.

Groundwater.   Groundwater is often saline. There are many of these aquifers in the basin. Therefore, the development of energy sources that release the water from these aquifers increases the salinity of the river. The salinity increase associated with mining coal is attributed primarily to leaching of coal spoil materials and the discharge of saline groundwaters. In northwestern Colorado the salinity of spoil-derived water ranges from approximately 3000 to 3900 mg L^{-1}. The salinity depends on the water residence time and the chemical and physical properties of the spoil. Saline water is also produced by oil and gas production. The salt increase due to energy activities can

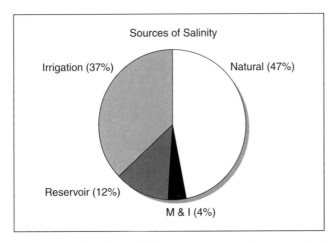

Figure 4.1.   Sources of salinity in the Colorado River basin. (From U.S. Department of the Interior, 1997.)
M & I = municipal and industrial; Reservoir = reservoir evaporation.

be attributed to consumptive use of good-quality water, mine dewatering, and if surface restoring is used, the leaching of spoil materials in a way similar to that of surface coal mining (U.S. Department of the Interior, 1997).

Another cause of increased salinity is the rise in water tables of saline aquifers. This is due to increased recharge or upward leakage from deeper aquifers, which not only causes land salinization but increases the seepage of saline groundwater into rivers and water courses, enhancing their salinization. Dams, weirs, and locks in regions where the groundwater is saline can generate or aggravate saline problems. These structures may create a head of water of several meters which would cause saline groundwater to flow into the rivers downstream from such structures. Furthermore, the increased water level will raise the water table levels in surrounding regions, leading to expansion of soil salinization problems. These effects may be due to the storage of large amounts of water in reservoirs behind the dams.

Runoff.   Salinity may also be increased by runoff during snowmelt and thunderstorms. Analyses of Green River water near the Green River, Utah station indicate that salinity remains unusually high during peak flows associated with snowmelt runoff events. Salt and sediment yields depend on storm period, landforms, and the soluble mineral content of the geologic formation. In the Price River basin, the highest salt and sediment concentrations occur in the first stream-flow event following a long period of no discharge. Yields occurring after initial flushing of the channel are similar to those found in the surrounding watershed soils (U.S. Department of the Interior, 1997).

An average of about 200 million acre-feet of water per year is provided by precipitation in the Colorado River basin. All but about 18 million acre-feet of this is returned to the atmosphere by evapotranspiration (U.S. Department of the Interior, 1997). Most of the stream flows originate in high forested areas where heavy snowpacks accumulate and evapotranspiration is low. A small amount of runoff originates at the lower altitudes, primarily from infrequent storms. Approximately two-thirds of the runoff is produced from about 6% of the basin area. Storm flows fluctuate widely from year to year and season to season because of variation in precipitation and the fact that numerous reservoirs have been constructed to make water available for local needs, local exports, and downstream obligations. The usable capacity of basin reservoirs is about 62 million acre-feet. The filling of these reservoirs causes a great fluctuation in stream flow.

Landforms.   Landforms affect sediment and the resulting salt yields. Three such major soil landforms are badlands, pediments, and alluvial valleys. Sediment yields may be as high as 15 tons/acre in badlands. Of the landform types, badlands produce the greatest amount of salt. Pediments are less erosive and therefore yield less salt to riverine systems; most of the salts have been leached from alluvial deposits. Thus erosion of their landforms yields less salt per volume of sediment than do the other two landforms, badlands and pediments. Alluvium channels incorporate both sediment and salt from the sloughed channel banks and salts from efflorescence at alluvial bedrock contacts (U.S. Department of the Interior, 1997).

Humanmade Sources of Salinity
The development of water resources has contributed to the increased salinity of the Colorado River in two major ways: (1) the addition of salts from water use, and (2) the consumption (depletion) of water. The combined effects of water use and consumption have had a significant impact on the salinity of the Colorado River basin. The damaging effects of these high levels of salinity prompted the basin states and the federal government to adapt salinity standards and an implementation plan to limit further increases in salinity. Irrigated lands from the Upper Colorado River Basin contribute about 3.4 million tons of salt per year, 37% of the salinity of the river. Irrigation increases salinity by consuming water and by dissolving salts from underlying saline soils and geologic formations, usually marine shales. The salts are mobilized by deep percolation found naturally in these soils, especially if the lands are overirrigated. Through salinity control practices, these contributions to the river system can be greatly limited, helping maximize future beneficial uses of the river.

Irrigation Effects.   Irrigation began in the Upper Basin in about 1860. The purchase of land from Native Americans in 1873 hastened irrigation practices, and by 1905, 800,000 acres were irrigated. Between 1905 and 1920, irrigated lands increased at a rapid rate, and by 1920, 1.4 million acres were being irrigated. It was reported that 1.6 million acres were irrigated in 1965. Since that time, irrigation has decreased somewhat, and in 1985 it was estimated that 1.4 million acres were irrigated in the Upper Basin.

Due to the difficulties of diverting water from the Colorado River, with its widely fluctuating flows, irrigation was developed slowly in the Lower Basin. The development of the Gila area began in 1875 and of the Palo Verde area in 1879. The Boulder Canyon Project was constructed in the 1930s, and other downstream projects since that time have provided continual expansion of irrigated areas. Twenty-one thousand eight hundred acres were added in 1970 by private pumping either directly from the

Colorado River or from wells in the floodplain. In 1980, 400,000 acres were being irrigated along the Colorado River main stream. Total irrigated land in the entire Lower Basin was about 1.5 million acres.

Irrigation is the major humanmade source of salinity. Annual salt pickup from irrigation averaged about 2 tons/acre above Hoover Dam. Below Hoover Dam, the average annual salt pickup from irrigation is about 0.5 ton/acre after the initial irrigating period.

Municipal and Industrial Sources.   Municipal and industrial sources of salinity are usually about 1% of the basin salt load, or 4% of the salinity. Iorns et al. (1965) estimated that municipal and industrial uses increased salinity by about 100 tons per 1000 people in the basin. The population reported in 1985 for the Upper Basin was 645,000 people. At the present rate of population increase, municipal and industrial sources will be increasing salinity by approximately 133,000 tons per year by the year 2010. However, it should be noted that most municipal wastes are relatively low in salt concentration compared with natural, industrial, and agricultural sources. In the Lower Basin, the Central Arizona Project is the last major development to deplete water from the lower Colorado River, approximately 1.5 million acre-feet per year (U.S. Department of the Interior, 1997).

Agricultural Sources.   Agriculture use is the single largest source of depletion in the Colorado River. Exports, reservoir evaporation, and municipal and industrial uses also account for lesser but significant depletion; the depletion of water or the reduction of volume flow has a significant effect. Since most of the exports from the Upper Basin are made from regions where the salinity concentrations are very low, loss of this high-quality water results in the remaining flows downstream becoming more concentrated. In the period 1941–1972, the average export was about 360,000 acre-ft per year. Completion of large projects has resulted in increased exports, to about 669,000 acre-ft per year.

Nearly half of the salinity in the Colorado River system is from natural sources. Salinity comes from saline springs, erosion of saline geologic formations, runoff from agriculture, reservoir evaporation, and municipal and industrial sources (Figure 4.1). The combination of the effects of water use and consumption have brought about significant salinity impacts in the Colorado River basin. Plans are now under way to limit further increases in the salinity (U.S. Department of the Interior, 1997).

Agriculture is the greatest contributor to increased salinity. For example: Irrigated lands in the Upper Colorado River Basin contributed about 3.4 million tons of salt per year, 37% of the salinity of the river. Increased salinity is caused by the consumption of water and the underlying salinity of the soils and geologic formations, especially marine shales. The use of underground irrigation should greatly reduce this consumption of water by evaporation and hence increased salinity from this cause. Many diffuse natural sources of salinity occur in the river basin, and these sources naturally increase the salinity of the water (U.S. Department of the Interior, 1997).

The single largest source of depletion is agriculture. Exports, reservoir evaporation, and municipal and industrial uses account for a less significant depletion. The fact that most Upper Basin exports are made at higher elevations, where salinity of the water is low, results in the remaining downstream flows becoming more concentrated.

Water export from the Upper Basin during 1941 to 1972 was about 360,000 acre-ft per year. Completion of large projects such as the Colorado Big Thompson, Duchesne Tunnel, Roberts Tunnel, and some more recent projects will export about 669,000 acre-ft of water per year (U.S. Department of the Interior, 1997).

Water from the Colorado system and its salinity concentration are very important to the state of Colorado. Recently, the state has made a compromise with the water authorities by placing underground irrigation in the agricultural lands, thus saving about half the water distributed in these lands which was formerly lost by evaporation. This has allowed California, particularly Los Angeles, to solve its problems temporarily and has not deprived the other states in the Colorado system of a larger amount of water, which might have been the case if this had not taken place.

Salinity has several bad effects on agricultural uses. It limits the types of crops that may be irrigated, reduces crop yields as salinity increases, and increases the salinity level in the root zones of soils. An increase of salt in the soil limits the types of plants that can be grown, and salinity above threshhold levels impairs crop yield. Irrigation water that has a high percentage of sodium ions affects the soil structure and has an adverse effect on crop production. The primary way to deal with increased salinity in soils is to switch to salt-tolerant crops or to apply more irrigation water and leach out excess salts.

Water for Mexico
To meet the quality and quantity of water to be supplied to Mexico from the United States, the Colorado River Basin Salinity Control Act of 1974 (Public Law 93-320) was enacted to provide Mexico with the quantity and quality of water specified in the agreement. To do this, the Coachella Canal was lined. It was estimated that this would save 141,000 acre-ft of water lost each year from this reach of the canal. Public Law 93-320 also authorized the construction, operation, and maintenance of the protective and regulatory pumping unit (PRPU) by the Bureau of Reclamation to manage and conserve U.S. groundwater for the benefit of the United States and for delivery to Mexico. It was estimated that pumping by the 35 wells of the PRPU would provide 125,000 acre-ft of water per year. This, combined with 15,000 acre-ft of water from wasteways and drains in the Yuma Valley, would furnish 140,000 acre-ft of water of the required 1.5 million acre-feet that must be delivered annually to Mexico. In addition, 35,000 acre-ft could be withdrawn by private wells and/or 242 wells for use on private land, to equal the 160,000-acre-ft limit for pumping in the 5-mile zone. Currently, wasteways and drains in the Yuma Valley exceed 15,000 acre-ft per year. If these drains and wasteways could be diminished and future wells added, the approximately 140,000 acre-ft could be delivered to the Southerly International Boundary at all times. In addition, approximately 23,500 acres of private, state, and state-leased lands have been aquired within the 5-mile zone of Mexico. The purpose of this was to limit development and hence preserve the groundwater, which was estimated to be about 160,000 acre-ft per year. These were the conservation measures under Title I (U.S. Department of the Interior, 1997).

Under Title II, the Bureau of Land Management was to reclaim and to diminish the amount of salt emitted from these lands. Twenty areas of land, known as *units*, were set up to manage the salt concentration of the river (U.S. Department of the Interior, 1997).

In 1940, the Bureau of Reclamation reclaimed the river by building the Welton–Mohawk Project, which added a drainage system to collect sump water and carry it away. Formerly, the drain emptied into the Colorado River just above the Mexican border. Due to the Welton–Mohawk drain, the Colorado River became much saltier and the salinity of the river soared to 1500 ppm at the Mexican border. This area is the most important agricultural area for Mexico. As a result, the Mexicans complained bitterly. The president of Mexico threatened to drag the United States before the World Court in the Hague. President Richard Nixon sent experts to study the conditions, and a Salinity Control Treaty was signed within a few months. In the 1970s, Congress authorized about $1 billion dollars to find engineering solutions to the Colorado salinity problem so that this area could be farmed and the quality of water to Mexico could be greatly improved.

The Yuma plant was authorized to remove sufficient salt from the Colorado River to allow the United States to comply with the treaty with Mexico. It was to cost around $300 per acre-foot of water. The salinity problem was not only that in the Colorado River but also in the San Joaquin Valley in California, a very important agricultural area. The salinity of the water has continued to be a very serious problem in areas to which the Colorado River was diverted, as well as in other rivers. Not only does this water contain salt, but selenium and pesticides are in the water, making it unusable for many purposes.

SUMMARY

Hardness.   The water below Lake Mead would be classified as very hard. After a certain degree of increase in hardness, water is unusable for many purposes.

Boiler feed and coolant water comprise a major portion of water use to industry in the basin. The quality of boiler feedwater is a large factor in the rate of scale formation on heated surfaces, degree of corrosion of the system, and quality of steam produced. In coolant water systems, resistance to slime formation and corrosion is affected by mineral quality. The quality of water in boilers and cooling systems is maintained by reducing mineral content to that of relatively good water quality.

Salinity.   The Colorado River is naturally a very saline river. Over 9 million tons of salt is carried down the river every year. The main effect of high salinity is economic. Therefore, the U.S. Geological Survey monitors the flow and salinity of river systems continuously throughout a network of 20 gauging stations. As development continues, more salinity control will be necessary. In 1997 it was estimated that measures were in place to control the salinity of 621,400 tons of salt. By the year 2015 it is estimated they would have to control 1,476,600 tons; that is, more than twice the present control measures would have to be in place.

It is believed that natural sources contribute the largest amount of salt to the Colorado River. The natural salt load of the Colorado River at Lees Ferry, Arizona has been estimated to be about 5.3 million tons per year. Natural point sources are saline springs. For example, it is estimated that the Glenwood–Dottsero Springs

Unit contributes 440,000 tons/yr and that the Paradox Valley Unit contributes 205,000 tons/yr. Natural sources contribute about two-thirds of the average annual salt load passing over Hoover Dam. In the lower part of the basin, mineral springs contribute more salt to the Colorado River than does any other salinity source. Groundwater is often saline.

Saline water is also produced by oil and gas production. Salt increases due to energy activities can be attributed to consumptive use of good-quality water, mine dewatering, and if surface restoration is used, the leaching of spoil materials in a way similar to that of surface coal mining. Another cause of increased salinity is the rising of water tables of saline aquifers. This is due to increased recharge and upward leakage from deeper aquifers. Dams, wiers, and locks in regions where the groundwater is saline can generate or aggravate saline problems. These structures may create a head of water of several meters, which causes saline groundwater to flow into surface waters downstream from such structures. Furthermore, increased water levels will raise the water tables in surrounding regions, leading to an expansion of the salinization problem.

Salinity may be increased by runoff during snowmelt and thunderstorms. Landforms affect sediment and the resulting salt yields. Three major soil forms associated with these sediments and resulting salt yields are badlands, pediments, and alluvial valleys.

The primary sources of pollution are several, of which irrigation is most important. Municipal and industrial sources of salinity are usually about 1% of the basin salt load, or 4% of the salinity. It is estimated that municipal and industrial uses increase salinity by about 100 tons per 1000 people in the basin. In 1985 in the Upper Basin, the population was about 645,000 people.

More than half the salinity in the Colorado River system is from natural sources. Agriculture is the greatest contributor to increased salinity. Exports, reservoir evaporation, and municipal and industrial uses account for less significant depletion.

BIBLIOGRAPHY

Ghassemi, F., A. J. Jakeman, and H. A. Nix. 1995. Salinisation of land and water resources: human causes, extent, management, and case studies. CAB International, Wallingford, Berkshire, England.

Iorns, W. V., C. H. Hembree, and G. L. Oakland. 1965. Water resources of the upper Colorado River basin. Technical Report. U.S. Geological Survey Professional Paper 441.

U.S. Department of the Interior. 1997. Quality of water. Colorado River Basin Progress Report 18.

Williams, W. D. 1987. Salinization of rivers and streams: an important environmental hazard. Ambio 16: 180–185.

CHAPTER 5

Management of
Water Quality Problems

HISTORY OF THE DEVELOPMENT
OF THE COLORADO RIVER[1]

Law of the River

As new uses develop, the *law of the river* has had to evolve to provide not only certainty of flow, but flexibility. Legal doctrines may have lagged behind changes in public attitude, but they have not prevented water policy changes to serve changing values, which include a consideration of insular minorities and environmental quality (Ingram et al., 1991).

The fundamental doctrine of Western water law, the rule of "first in time, first in right," was conceived to make possible the building of projects that diverted water from mainstream rivers to land sometimes remote from the basin. The Prior Appropriation Doctrine prevented large upstream diversions that might drain the water supply. The water law was fashioned not only to allocate water and to facilitate development but also to cushion development from risk of shortages and drought. As the use of water for agricultural and urban demands grew, the Western states persuaded the federal government to secure new supplies through carryover storages and interbasin diversion projects (Ingram et al., 1991).

Management of Water. Many people thought that by managing water they could increase settlement of the West. Indeed, President Theodore Roosevelt, in a State of the Union Address, envisioned that the western half of the United States

---

[1] The information in this section is derived primarily from Reisner (1993) and U.S. Department of the Interior (1997).

would sustain a population greater than that of our entire country today if the waters that now run to waste were saved and used for irrigation. Although Congress attempted to make this a realization, it never happened. The Reclamation Act of 1902 provided loans to attract small farmers to these previously arid areas. When the farmers could not repay their debts, other legislation, beginning with a $20 million loan in 1910 and the Curtis Act of 1911 and others to follow, such as the 1939 Reclamation Project Act, made it easier to renegotiate reclamation contracts. According to the Congressional Report of 1914, the Reclamation Extensive Act was specifically intended to create an irrigated empire in the West. Reclamation laws not only made available new water for irrigation, but also reduced the risk involved in development (Ingram et al., 1991).

The Colorado River Compact of 1922 divided the Colorado River drainage basin into the Upper Basin, comprising the states of Colorado, New Mexico, Utah, and Wyoming, and the states of the Lower Basin: Arizona, California, and Nevada. In addition to providing equity between the Upper and Lower Basins, the major purposes of the compact were "to establish the relative importance of different beneficial uses of water; to promote interstate comity; to remove causes of present and future controversies; and to secure the expeditious agricultural and industrial development of the Colorado River basin, the storage of its waters, and the protection of life and property from floods." The negotiators of the compact assumed that they were dividing an average annual flow at Lees Ferry of 16 million acre-feet, providing 7.5 million acre-feet annually to each basin. Article III(b) of this contract provided to the Lower Basin grants for an additional right to be assumed, 1.5 million acre-feet, which would assume the 1 million acre-feet surplus. Article III(d) provided that the Upper Basin states would not cause the flow of Lees Ferry to be depleted by 75 million acre-feet for any consecutive 10-year period, and thus protect Lower Basin uses (Ingram et al., 1991).

The basin states provided that any amount granted to Mexico by treaty would be divided equally between the two basins. The Lower Basin asserted that it had the right to use any water not immediately put to use in the Upper Basin through a provision which declares that the Upper Basin states cannot withhold the delivery of water that cannot reasonably be applied to domestic or agricultural uses. The reciprocal duty of the Lower Basin was not to demand the delivery of water to the Lower Basin for use other than domestic and agricultural purposes. This statement was meaningless because this was the entire usage of water in the Lower Basin. California benefited primarily by Article III.

In contrast, the Upper Basin merely acquired the opportunity to use its entitlement, as opposed to obtaining firm rights for the use of 7.5 million acre-feet per year. As a result of this, the delegates gave high priority to the use of water for agricultural and domestic purposes. Other uses, such as power generation and navigation, were also provided in the compact, but had a lower priority. For example, Article IV(b) expressively states that the impounding and use of water for hydropower "shall be subservient to the use and consumption of such water for agricultural and domestic purposes and shall not interfere with or prevent the use for such domestic purposes," while Article IV(a) makes navigation "subservient" to the three other uses (Ingram et al., 1991).

Changes in the Law of the River over Time.   Just as the Grand Canyon is a living geological record, the law of the river can be measured by time. Time helps to define the law's various levels of command. The law adopts different time horizons, from the concept of the eternal to the frantic pace of an hour. A time horizon also helps to chart the ever-increasing complex of accelerations that are changing the law. For example, the law of the river speaks to the following time commands: (1) *eternity*, interbasin allocations and allocations among the basin states; (2) *indeterminate*, possible flow necessary to protect endangered species; (3) *decade*, Upper Basin delivery obligations; (4) *year*, Lake Powell fill obligations and the Lake Powell / Lake Mead balance; and (5) *hour*, WAPA Western Area Power Administration power generation decisions.

The law of the river includes the totality of compacts, international agreements, treaties, judicial decisions, statutes that speak directly to the river and to water allocation generally, and the administrative practices that influence allocation of the river. This expanded definition of the law encompasses both immutable commands and mutable principles (Ingram et al., 1991).

Water Supply

By the 1960s, having sufficient water in the Colorado River to satisfy all the demands had become a great problem. On August 18, 1965, Randy Ritter, the Bureau's resident expert on the Colorado River, forwarded a letter to Commissioner Floyd Dominy. It stated that there was not enough water in the Colorado River to permit the Upper Basin to use fully its apportionment of 7.5 million acre-feet and still meet its Compact obligations to deliver water at Lees Ferry. Tipton, a consulting hydrologist with the Bureau of Reclamation, felt that the estimate of the river flow should be set no higher than 15 million acre-feet. In that case, if one divided the shortages equally between the two basins, each would be left with 6.3 million acre-feet. After deducting another 1.5 million acre-feet, which they had to give to Mexico, the figure was so low that it threw seven states into panic. It was evident that the future of the Colorado River basin depended on "importing water to augment the dependable supply in the basin." Ritter suggested that as a minimum, the Central Arizona Project (CAP) legislation pending before Congress be rewritten to contain a conditional authorization of an import plan of at least 2.5 million acre-feet (Reisner, 1993).

President Eisenhower decided to do nothing until the Central Arizona Project was given final say. This had not happened, so legal battles had been entered to determine who had the right to what water. Since the diversion from the Klamath River seemed impossible both politically and actually, the plans turned to raising the flow of the Colorado River to Tucson and Phoenix, but this would involve a pump lift to 1200 ft. (365.76 m). Pumping irrigation water there would be like taking it out of the Hudson River and lifting it over the World Trade Center in order to water lawns on Long Island. The purpose of this was to secure water for irrigation projects for dying farmlands between Phoenix and Tucson. There were no sites for big dams left in Arizona besides the Gila River system, which did not have nearly enough water to develop the kind of power the Bureau had in mind.

Central Arizona Project.   In 1965 the Central Arizona Project was authorized, but it had been pruned somewhat. However, a new aqueduct to Las Vegas was also authorized. CAP authorized something called a *development fund,* a receptacle for revenue from the power dams that in the future would help finance the augmented scheme everyone knew would be needed. Legislation would, in other words, authorize the projects that would ensure the Colorado's early exhaustion. The Central Arizona Project, signed into law by Lyndon Johnson on September 30, 1968, was the most expensive single authorization in history. Besides the Central Arizona Project, it authorized Hooker Dam in the Gila Wilderness of New Mexico, the aqueduct from Lake Mead to Las Vegas, the Dixie Project in Utah, and the Uinta Unit of the General Utah Project, the first piece of the water diversion scheme, which promised to be nearly as grand as CAP. It also authorized the San Miguel, Dallas Creek, West Divide, Dolores, and Animas–Lapada Projects in Colorado. It authorized the Lower Colorado Development Fund, still penniless, to build an augmented project that had not yet been defined, let alone approved. Almost unnoticed alongside everything else, the bill authorized delivery to Mexico of 1.5 million acre-feet of tolerably sweet or fresh water. This was a national responsibility, whatever it meant. Loosely interpreted, it might mean a pipeline from Lake Superior to Mexico (Committee to Review Glen Canyon Environmental Studies, 1991).

The great demand for irrigation water was a serious problem for the Colorado. Many farms and towns were drawing water from the Ogillala Aquifer, which was dropping rapidly due to overuse. This is a large nonrechargeable aquifer that underlies New Mexico and Arizona. Various other projects, which were very expensive, were developed to supply the needs of the growing West.

Dams.   Various tributaries are also controlled by dams. The flow of the San Juan River is controlled by the Navajo Dam, the Green River by the Fontenelle and Flaming Gorge Dams, and the Gunnison River by the Wayne N. Aspinall Unit Dam. Glen Canyon Dam is the only major dam on the main stem of the Colorado River above Lees Ferry, but it controls almost all the flow leaving the Upper Basin. Lake Mead, formed by the Hoover Dam, supplies most of the storage and regulation in the Lower Colorado Basin. It provides water for irrigation, municipal, and industrial use, power generation, flood control, recreation, and many other beneficial uses. Lake Mohave, the reservoir formed by Davis Dam, which is located downstream from the Hoover Dam, backs water upstream at high stages about 67 miles (107.83 km) to the tailface of the Hoover power plant. Storage in Lake Mohave is used for some regulation of releases from the Hoover Dam, for meeting treaty requirements with Mexico, and for developing a power head for the production of electrical energy at the Davis Power Plant. The river flows through a natural channel for about 10 miles (16.1 km) below the Davis Dam, at which point the river enters the broad Mohave Valley, 33 miles (53.11 km) above the upper end of Lake Havasu. Lake Havasu backs up behind Parker Dam for about 45 miles (72.42 km) and serves as a forebay from which the Metropolitan Water District of Southern California pumps water into the Colorado River Aqueduct. Lake Havasu also serves as a forebay for the Central Arizona Project Havasu Pumping Plant and pumps water into the Hayden Rhodes Aqueduct.

Alamo Dam and Reservoir on the Bill Williams River is used to control floods originating above and below Alamo Dam. Imperial Dam, located some 150 miles (241.4 km) downstream from Parker Dam, is the major diversion structure for irrigation projects in the Imperial Valley and Yuma areas. It diverts water on the west bank to the All American Canal, which delivers water to the Yuma Project in Arizona and California and to the Imperial and Coachella Valleys in California. It diverts on the east bank to the Gila Gravity Main Canal. The Senator Wash Dam, an offstream storage facility, also affords regulation in the vicinity of Imperial Dam and assists in the delivery of water to Mexico. Morelos Dam, located just below the Northerly International Boundary with Mexico, is the last dam on the Colorado River. This small diversion dam diverts water into the Alamo Canal, which delivers water to northern Mexico (U.S. Department of the Interior, 1997).

Summary of the Salt Load Distribution

Salinity projections for 1970 conditions of limited development were made on the basis that water resources development then in operation and 1970 water-use patterns would hold for a repetition of the 1942–1961 hydrological record. The 1970 projections reflected the effects of evaporation losses from Lake Powell, operating at normal levels. Since Lake Powell had not yet reached normal storage levels, evaporation losses were less than expected average losses, and average salinity levels at downstream points were correspondingly lower than projected. For the 1980 conditions of limited development, it was assumed that no new water resource development would be placed in operation but that those projects then under construction would be completed as planned. It was assumed that all construction would be completed by 1980 and that 2010 conditions of water use would remain the same as those for 1980. In the past, salt loadings had been the dominant factor affecting salinity concentrations and contributed about three-fourths of the average salinity concentrations at Hoover Dam under 1960 conditions. In the future, however, increased salinity levels would result primarily from flow depletion caused by out-of-basin export, reservoir evaporation, and consumptive use of water for municipal, industrial, and agricultural purposes.

In the next 40 years, it was projected that there would be a relatively constant average salt load; however, the water flow drop would be substantial. Over 80% of future increases in salinity concentrations at Hoover Dam would be due to increased flow depletions. Between 1960 and 2010, over three-fourths of the projected salinity increase would be the result of increases in reservoir evaporation brought about by filling of the major storage reservoirs completed since 1960 and the increase in consumptive use brought about by the expansion of irrigation. As salinity increased, certain water uses would be impaired. It was expected that future salinity concentrations in the Colorado River basin would be below threshhold limits for in-stream uses such as recreation, hydroelectric generation, and propogation of aquatic life. Only marginal impairment of these uses was anticipated. In the lower Colorado River, salinity concentrations were above threshhold limits for municipal, industrial, and agricultural uses. Some impairment of these uses was occurring, and it was believed that future increases in salinity would increase this adverse effect.

Control Projects

Area Covered by 1960 Project. The geographical area covered by the 1960 project included the entire Colorado River basin and Southern California Water Service Area. The Colorado River is also utilized by Mexico. The Colorado River Project included a large range of studies, including hydrology; chemistry; mathematics; computer science; soil science; geology; civil, sanitary, and agricultural engineering; and economics.

The Colorado River Water Quality Control Project was established in 1960. The policy was then spelled out over 14 years in a series of legislative acts described as the Federal Water Pollution Control Act. As amended, 33USC 466SEQ Section 10D of this act directs the Secretary of Interior, whenever requested by any state water pollution control agency, if the request refers to pollution of water that is endangering the health and welfare of people in a state other than that in which the source of pollution originated, to call a conference with other states that may be affected adversely by such pollution.

Salinity was also a problem to users in Mexico and in limited areas of the Upper Colorado River Basin. Salinity concentrations are affected primarily by two basic processes: *salt loading*, the addition of mineral salts from various natural and humanmade sources; and *salt concentrating*, the loss of water from the system through evaporation, transporation, and out-of-the-basin export. Using data from 1942 to 1961, the salinity concentration at Hoover Dam was estimated to average about 700 and 760 mg L^{-1} under 1960 and 1970 conditions, respectively. If no salinity controls were implemented, the salinity concentrations at Hoover Dam were estimated to increase to about 880 mg L^{-1} in 1980 and 980 mg L^{-1} in 2010. Comparable figures at Imperial Dam were 760 and 870 mg L^{-1} under 1960 and 1970 conditions and 1060 and 1220 mg L^{-1} under 1980 and 2010 conditions. If water resources in the future were limited to completion of projects that were currently under construction, it was estimated that the average annual salinity concentrations for 1980 and subsequent years would increase to only about 800 mg L^{-1} at Hoover Dam and 920 mg L^{-1} at Imperial Dam. Salt concentration would be 800 mg L^{-1} instead of 870 mg L^{-1} at Hoover Dam and 920 mg L^{-1} instead of 1060 to 1220 mg L^{-1} under 1980 and 2010 conditions.

Compacts and Control Projects Related to Salinity. The Conference Concerned with Pollution of Interstate Waters of the Colorado River was held on January 13, 1960. The conference was requested by six of the seven basin states. One meaning of the conference was that salinity (total dissolved solids) was the most serious water quality problem in the Colorado River basin. Average salinity concentrations ranged from less than 50 mg L^{-1} in the high mountain headwaters to about 865 mg L^{-1} at Imperial Dam, the last point of major water diversion in the United States.

Salinity affects the water supply for a population exceeding 10 million people and for 800,000 irrigated acres in the Lower Colorado River basin and Southern California Water Service Area. Salinity has also been a problem to users in Mexico and limited areas of the Upper Colorado River Basin (Reisner, 1993).

Various alternatives suggested by the 1960 Conference included augmentation of the basin water supply. This would be achieved by importation of demineralized seawater, importation of fresh water from other basins, or utilization of weather-modification techniques to increase precipitation and runoff.

Reduction of the salt load could be achieved by impoundment and evaporation of saline waters from point sources, diversion of runoff in streams around the area of high salt pickup, improved irrigation drainage practices (e.g., underwater irrigation), improved irrigation conveyance abilities, desalination of saline discharges from natural and humanmade sources, and desalination of water supplies at points of use, with appropriate disposal of the waste brine.

A basinwide salt load reduction program was developed that would reduce the salt load contributions by five large natural sources and 12 irrigated areas, totaling 600,000 acres. Once implemented, it was estimated that this program would reduce the average salinity concentration at Hoover Dam by about 250 mg L^{-1} in 1980, increasing to 275 mg L^{-1} in 2010 (U.S. Department of the Interior, 1997. Quality of Water, Colorado River Basin Progress Rpt. No. 18).

The program would have an estimated average annual cost of $7 million in 1980 and $13 million in 2010. Such a program would limit salinity concentrations at Hoover Dam to approximately 1970 levels. The direct salinity control benefits of such a program were estimated at $11 million in 1980, increasing to $22 million in 2010 (in 1970 dollars) (U.S. Department of the Interior, 1997).

The recommendations of the conference were that a salinity policy be adopted for the Colorado River system that would have as its objective the maintenance of salinity concentrations at the low levels then found in the lower main stem. Second, specific water quality criteria standards would be adopted at key points throughout the basin by the appropriate states in accordance with the Federal Water Pollution Control Act. Such a program would have as its objective keeping the maximum mean salinity concentration at Imperial Dam below 1000 mg L^{-1}. Implementation of the recommendation policy and criteria would be accomplished by carrying out a basinwide salinity control program that included planned future development of the basin's water resources.

Cost of Increased Salinity. The salinity levels in the Hoover Dam are a key point in the Colorado system, as most points of use in the Lower Basin and Southern California Water Service Area relate directly to the salinity at the Hoover Dam. Salinity detriments are the sum of direct costs to water users and indirect penalty costs. In 1960 it was estimated that the annual economic impact of salinity was $9.5 million. If water resource development were to proceed as proposed and no salinity controls were implemented, it was estimated that the average annual detriment in 1970 dollars would increase to $27.7 million in 1980 and $50.5 million in 2010. If future water resource development were limited to projects under construction in 1960, estimated annual economic detriments would increase to $21 million in 1980 and $29 million in 2010.

Possibilities for minimizing and controlling salinity in the Colorado River Basin may be divided into two categories: water-phase and salt-phase control measures. *Water-phase measures* seek to reduce salinity concentrations by augmenting the water supply. *Salt-phase measures* seek to reduce salt input into the river system. Various factors, such as economic feasibility, lack of research, and legal and institutional constraints, limit the present application of water-phase and salt-phase control measures. The most practical means of augmenting the basin water supply include importing water from other basins, importing demineralized seawater, and utilizing weather-modification techniques to increase precipitation and runoff in the basin.

Practical means of reducing salt loads include impoundment and evaporation of point-source discharges, diversion of runoff in streams around the area, salt pickup, improved irrigation and drainage practices, improved irrigation conveyance facilities, desalination of saline discharges from natural and humanmade sources, and desalination of water supplies at points of use. These measures can be implemented in a variety of ways and in several different combinations.

Several attempts have been made to control the salinity, and all are expensive. But the thought of doing nothing is untenable. The second approach would eliminate economic or water resource development that is expected to produce an increase in salt loads or stream flow depletion. Such an approach would minimize future increases in the economic impact of salinity and possibly eliminate the need for salinity control facilities (and cause deviation in the growth of the region's economy). Projections of future salinity levels and associated salinity detriments for this approach have been discussed and they are untenable. The third approach, calling for the construction of salinity control works would allow water resource development to proceed.

Laws and Regulation
Laws Governing Colorado That Affected Salinity. In addition to state laws that provide for interstate control of water, use of water in the Colorado River system is governed principally by four documents: the Colorado River Compact (signed in 1922), the Mexican Water Treaty (signed in 1944), the Upper Colorado River Basin Compact (signed in 1948), and the Supreme Court Decree of 1964 in *Arizona v. California*. The Colorado River Compact provides 7,500,000 acre-ft of water annually from the Colorado system for both the Upper and Lower Basins. It further establishes the obligation of Colorado, New Mexico, Utah, and Wyoming, designated as states of the upper division, not to cause the flow of the river at Lees Ferry to be depleted below an aggregate of 75 million acre-feet for any period of 10 consecutive years.

In accordance with the Upper Colorado River Basin Compact, Arizona is granted consumptive use of 50,000 acre-ft of water per year, and the other Upper Basin States are each appropriated a percentage of the remaining consumptive use as follows: Colorado, 51.75%; New Mexico, 11.25%; Utah, 23%; and Wyoming, 14%. Of the 7,500,000 acre-ft of Colorado River water entering the Lower Basin, the states of Arizona and Nevada are appropriated 2,800,000 acre-ft for Arizona and 300,000 acre-ft for Nevada.

For the lower division, apportionment was divided among the Lower Basin States—Arizona, California, and New Mexico—by decree of the U.S. Supreme Court, which stated in 1964 that the appropriation was accomplished by the Boulder Canyon Project Act of 1929. Colorado River main-stem water is available in sufficient quantity to satisfy the 7,500,000 acre-ft of annual consumption. Use in the three Lower Basin states is apportioned at 2,800,000, 300,000, and 4,400,000 acre-ft, respectively for Arizona, New Mexico, and California. The major uses of water within the basin are for agricultural, municipal, and industrial purposes. Most of the water in the Lower Basin is used for irrigated agriculture. The remaining small amount is used principally for municipal and industrial uses. Approximately three-fourths of the 7 million acre-feet of water consumed each year is depleted by agricultural uses. In other words, 7 million acre-feet is lost to the system by evaporation during agriculture use. Minor quantities are consumed by hydroelectric and thermal production, recreation, fish and wildlife, rural domestic needs, and livestock.

In the urban areas of the basin, municipal and industrial uses are increasing significantly, due to the rapid rate of population growth. One of the largest losses from the streams is by surface evaporation from storage reservoirs. Over 2 million acre-feet is estimated to evaporate annually from the lakes and reservoirs of the basin.

Colorado Compact, 1922: Division of Colorado River, Upper and Lower Basins. The Colorado River Compact of 1922 (45 Stat. 1057) established a division point of the Colorado River at Lees Ferry, Arizona to separate the Colorado River basin into Upper and Lower Basins for legal, political, institutional, and hydrological purposes. Lees Ferry is located about 1 mile above the confluence of the Paria River and approximately 17 miles downstream from Glen Canyon Dam (U.S. Department of the Interior, 1997).

Characteristics of the Upper Basin. The Upper Basin comprises about 45% of the drainage area of the Colorado River basin. Weather variation is extreme, and ranging from hot and arid in desert areas to cold and humid in the mountain ranges. In low elevations or in the rain shadow of coastal mountain ranges, desert areas may receive as little as 6 in. of precipitation annually, while the mountain areas may receive more than 60 in. Temperatures range from temperate, affording only a 90-day growing season in the mountain meadows of Colorado and Wyoming, to semitropical conditions with year-round cropping in the Yuma–Phoenix area. On a given day, both the high and low temperature extremes for the continental United States may occur within the basin. In the Southern California Water Service Area, the climate around the Salton Sea is hot and arid, whereas the climate of the coastal metropolitan areas is moderate.

Problems Associated with Water Quality Standards
The Water Quality Act of 1965 provided that all states establish water quality standards for all interstate streams. Subsequently, the seven basin states developed water quality standards for the Colorado River. The standards established for the states did not include numerical salinity standards. The Secretary of Interior approved the water quality standards provided that numerical salinity standards would be established at such future time as sufficient information had been developed to provide the basis for workable, equitable, and enforceable salinity standards. The states are still faced with establishing suitable salinity standards in compliance with the Water Quality Act of 1965.

It should be noted that water quality and water quantity considerations are generally under the jurisdiction of different agencies at both the state and federal levels. This split jurisdiction poses coordination problems to all interests affected by a salinity control program. Legal and institutional arrangements would also place constraints on the means available to finance salinity control programs. Salinity control below the Hoover Dam is a possible practical approach toward minimizing the economic impact of salinity and should receive further consideration in the formation of a basinwide salinity control program.

Alterations in salinity concentration, resulting from factors such as seasonal changes in stream flow and water use, occur throughout the basin. The possible magnitude of such fluctuations and their adverse impact, however, would indicate the need for more positive means of minimizing peak concentrations. Other water quality simulation capabilities will therefore need to be refined before the effectiveness of control measures can be evaluated.

FUTURE

Causes of Future Increases in Salinity

As stated above, in 1960 it was estimated that evaporation and irrigation would account for almost three-fourths of the salinity increase between 1960 and 2010. This increase in salinity was expected to greatly decrease the value of the land for agricultural purposes. In 1960 the annual economic detriments caused by salinity were estimated to total $16 million. It was thought that if development proceeded as proposed and no salinity controls were implemented, it is estimated that the average annual economic detriment would increase to $28 million in 1980 and $51 million in 2010. More than 80% of the total future economic detriments was estimated to be incurred by irrigated agriculture located in the Lower Basin and Southern California Water Service Area and by associated regional economy. About two-thirds of these detriments will be incurred directly by irrigation water users and the remainder will be incurred indirectly by other industries associated with agriculture.

Future Effects on Population and Economy of the Colorado River Basin

The Colorado River Basin on the whole is sparsely populated. In 1965 the estimated population was nearly 2.25 million people, with an average density of about 9 persons per square mile compared to a national average of 64 persons per square mile. Eighty-five percent of the population lived in the Lower Basin. About 70% of the Lower Basin population resided in the metropolitan areas of Las Vegas, Nevada and Phoenix and Tucson, Arizona. It was estimated that the population of the Colorado River Basin would triple by 2010. The Southern California Water Service Area serviced an estimated 11 million persons in 1965, in contrast to the 2.25 million persons in the Colorado River basin. Most of the population was concentrated in the Los Angeles–San Diego area.

The economy of the basin is based on manufacturing, irrigated agriculture, mining, forestry, oil and gas production, livestock, and tourism. The last two decades the economy of the Lower Basin has experienced a significant transition from an agricultural mining base to a manufacturing service base. Growth in the manufacturing sectors has been one of the major factors in the overall economic growth of the Lower Basin. Important manufacturing categories are electrical equipment, aircraft and parts, primary metal industry, food and kindred products, printing and publishing, and chemicals. Agriculture continues to play an important part in southern California's economy, amid the vast and growing industrial and commercial activities. Manufacturing is the most important industrial activity and principally includes the production of transportation equipment, machinery, food and kindred products, and apparel. Agriculture accounts for about 3% of the total employment, manufacturing an estimated 30%, and trades and services approximately 42%.

Factors Affecting Salt Concentration

Stream-Flow Depletion. Stream-flow depletion contributes significantly to salinity. Consumptive use of water for irrigation is the largest depletion source. Consumptive use of water for municipal and industrial purposes accounts for a much smaller depletion. Evaporation from reservoirs and stream surfaces produces large depletions. Phreatophytes cause significant loss of water by evapotranspiration, especially in the Lower Basin below Hoover Dam.

Diversion of Water. Diversion of water out of the basin, particularly in the Upper Basin, is a problem. The water diverted is high in quality and low in salt content. Thus while these diversions remove substantial quantities of water from the basin, they remove only a small portion of the salt load. Only about 47% of the average salinity concentrations for the 20-year period 1941–1961 were attributed to natural factors. This evaluation indicates that the salinity concentrations would have averaged only 334 mg L^{-1} at the Hoover Dam location under natural conditions for 1941 to 1961.

Potential Impact of Increased Salinity

Initial investigations conducted on the potential impact of future salinity levels revealed that only small effects on water uses could be anticipated in the Upper Basin. Subsequent investigations were limited to three main study areas: the lower main stem and Gila areas, the Lower Basin, and the area encompassing the Southern California Water Service Area. The boundaries of these study areas follow political rather than hydrological boundaries. Although significant economic effects are known to occur in Mexico, lack of data preclude their inclusion. The effect of salinity increases on the basin water indicate that adverse physical effects would essentially be limited to municipal, industrial, and agricultural uses. These uses comprise the major utilization of water supplies.

Management Objectives

The 1960 study proposed that at least three possible management objectives be considered: Salinity controls could be implemented to control the salinity at a given level, salinity could also be maintained at a level that minimizes its total economic impact, and salinity could be maintained at that low level at which the total economic impact of salinity would equal the impact that would be produced if no action were taken. A program should be developed for salinity control.

The three alternatives for controlling salt load were reduction programs, four flow-augmented programs, and one program to demineralize water supplies at the point of use. The three salt load reduction programs utilized measures such as (1) desalination or impoundment and evaporation of mineral spring discharges, irrigation return flows, and saline tributary flows; (2) diversion of streams; and (3) improvement of irrigation practices and facilities. Such programs would reduce the salt load approximately 3 million tons annually and reduce the average annual salinity concentration at Hoover Dam to about 200 to 300 mg L^{-1}. The four flow augmentation programs were based on three potential sources of water: (1) increased precipitation through weather modification, (2) interbasin transfer of water, and (3) importation of demineralized seawater. The flow provided by these programs would be 1.7 to 5.9 million acre-feet annually, and this would result in an annual salinity concentration reduction at Hoover Dam of between 100 and 300 mg L^{-1}. The last alternative program evaluated would utilize desalinated water supplies diverted to southern California. This would minimize the adverse impact of salinity on the Southern California Water Service Area.

Including amortized construction costs, operating costs, and maintenance, costs were estimated to range from $3 million to $177 million annually. The worth of the total program cost for each alternative from 1975 to 2010 would range from $30 million to $1570 million. It was estimated that if nothing were done, the average salinity concentrations at Hoover Dam would be 876 and 990 mg L^{-1} in 1980 and 2010, respectively.

Comparison of the Alternative Salinity Control Programs.   The phase implementation of the salt load reduction program was selected as the lowest-cost alternative for achieving basinwide management and control of salinity. However, the combination of flow augmentation with the salt load reduction program was thought to be the optimal approach. This program would be designed to reduce the salt load contribution by five large natural sources and 12 irrigated areas, totaling 600,000 acres. The five natural sources contributed about 14% of the basin's salt load. All of the irrigated areas selected exhibited high salt pickup by return flow of about 3 to 6 tons/acre per year. This acreage comprised only about 20% of the basin's irrigated load from irrigated sources above Hoover Dam.

For the 17 projects the average annual cost, including operation, maintenance, and amortized construction costs, was estimated. For five single-purpose salt load reduction projects, all costs were assigned to salinity control. The improved irrigation projects would be multipurpose. It was estimated that they would produce various economic benefits of about the same magnitude as salinity control benefits, and for this reason only half of the cost of irrigation improvement was allotted to salinity control. Estimates of the change in stream flow depletions and salt load reductions were also prepared for each project. Five salt load reduction projects were estimated to remove an average of about 172,000 acre-feet per year from the river system above Hoover Dam. Of this amount, 130,000 acre-ft of demineralized water from the Blue Springs project would be available for use in central Arizona. Irrigation improvement projects would reduce nonbeneficial consumptive water use by an estimated average of 299,000 acre-ft per year.

Potential Effects and Costs of Salinity Control Programs
Benefits.   A salinity control program was projected to result in a net increase in available basin water supply of more than 250,000 acre-ft per year. The incremental

reductions in average salinity concentrations at Hoover Dam were estimated for each control project for the years 1980 and 2010 by using predicted changes in flow and salt load. The salinity reduction for each project was estimated to be greater in 2010 than in 1980.

Costs. At a given salinity level there is an economic cost associated with water use and a second economic cost associated with maintaining salinity concentrations at that level. The sum of these costs, defined as total salinity costs, can be determined for any time period and salinity level by proper manipulation of three factors: the salinity detriment functions, the salinity management cost functions, and the predicted future salinity concentrations with no control implemented. Salinity controls would have an effect on both water quality and economic objectives. Effecting water quality objectives has associated economic effects. Conversely, the selection of an economic objective will result in the selection of associated salinity levels. Knowledge of the interrelationships between economic and water quality effects is thus useful in the rational selection of salinity management objectives. In the 1960 study, economic and water quality effects associated with three salinity management objectives were determined: (1) to maintain salinity at a level that would minimize its total economic impact and achieve economic efficiency (the minimum-cost objective), (2) to maintain salinity concentrations at some specific level (the constant-salinity objective), and (3) to maintain salinity at some low level for which the total economic impact would equal the economic impact that would be produced if no action were taken.

Restricting Development. The study theorized that total salinity cost would be minimized by limiting development alternatives. The determination of the net economic benefits if the limited development approach were utilized was beyond the scope of the project.

The future economic impact of salinity will be great. Although implementing salinity controls will result in the availability of better quality water for various uses, and some of the economic impacts would shift from salinity detriments to salinity management costs, the total economic impact of salinity would not be reduced substantially. As a minimum, it was estimated that average annual total salinity costs would increase threefold between 1960 and 2010. Selection of the limited development alternative was expected to reduce the annual cost by only about 40% below the no-control alternative in the year 2010.

With no controls implemented, average annual salinity concentrations at Hoover Dam were predicted to increase between 1960 and 2010 by about 42%, or 293 mg L^{-1}. Distribution of salinity costs related to each alternative differs greatly. Distribution of costs may therefore be an important factor in selecting the alternative.

Extremes in the range of cost point out the basis for equity considerations that may enter into the selection of management objectives. If the no-cost-control alternative were selected, all salinity costs would essentially be borne by water users and by the regional economy in the Lower Basin and Southern California Water Service Area. In contrast, selection of the equal-costs alternative would redistribute a majority of the costs to investments in salinity control facilities in the drainage area upstream of Hoover Dam. Much of this investment would be for irrigation improvement in the Upper Basin, improvements that would produce substantial economic benefits in addition to salinity control benefits. The equity of these two extremes in cost distribution was seen to be vastly different.

Legal and Institutional Constraints.   The implementation of a basinwide salinity control program based on salt load reduction projects faced a number of legal and institutional constraints. One of the most formidable would be imposed by existing state water laws and regulations concerning water rights and beneficial uses. At the time of the study these laws did not recognize the utilization of water for quality control purposes as a beneficial use.

Improved irrigation efficiencies would reduce the amount of water required for diversion to a given farm or irrigation project. The effects of such a reduction in water use on perfected water rights was unclear and thus could cause legal problems. Such legal factors might affect the selection of control measures to be incorporated in a basinwide salinity management program.

It was felt that the limited development alternative would result in slight increases in average salinity concentrations. The total economic impact of salinity associated with each alternative evaluated varied over a limited range. The distribution of salinity costs related to each alternative differed greatly, so distribution costs might be an important factor in selecting among the alternatives.

Future Water Quality Objectives of Basin States

The range of possible problem solutions points out the need for rational selection by the basin states of objectives for future water quality and uses and the formulation of a basinwide salinity control plan to meet these objectives. It is recommended that a broad water quality objective be adopted by the basin interests which would require salinity concentrations to be maintained at or below present levels in the lower Colorado River. This objective would become part of the basic policy regarding the comprehensive planning and development of the basin's remaining water resources. It is recommended that appropriate Colorado River basin states take the steps necessary to establish a numerical objective for salinity concentration.

Evaluation of water quality effects of various salinity control alternatives has shown that either by implementing a basinwide salinity control program or by limiting water resource development, future salinity levels at Hoover Dam could be maintained at or below an annual concentration of 800 mg L^{-1}. A corresponding limit of 1000 mg L^{-1} at Imperial Dam could be achieved. A limit on average monthly concentrations is considered necessary to provide a more acceptable level of protection. To achieve compliance with the basic policy objectives to enhance water quality in the lower Colorado River will require that detailed salinity criteria be established at a number of key locations throughout the basin. By maintaining salinity levels at upstream locations below assigned levels, compliance with downstream criteria will be assured. Second, the criteria will provide a basis for optimum development of the resources of a given tributary, subbasin, or state. Complete basinwide salinity criteria should be established after careful consideration of the basin interest of such factors as existing salinity level, proposed water resource development, the feasibility of salinity control, water quality requirements for water uses, and the economic impact of salinity. It is recommended that a state and federal task group be established to carry out the necessary activities to develop detailed salinity criteria for key control points in the basin. This group should include representatives from federal, state, and local agencies.

Since the Water Quality Act of 1965 provided that all states establish water quality standards for all interstate streams, and this was not done for Colorado, the states are thus still faced with the task of establishing suitable salinity standards in compliance with the Water Quality Act of 1965. The lack of numerical salinity standards may be a constraint to the rational planning of water resource development and implementation of salinity control.

One important fact is that there is not a single entity with basinwide jurisdiction to direct and implement a salinity control program. In addition, water quality and water quantity considerations are generally under the jurisdiction of different agencies at both the state and federal levels. This split jurisdiction poses coordination problems to all interests affected by the salinity control program.

Detailed analyses have not been made for a potential means of financing the program. A cursory review of the program's available financing facilities similar to those contemplated indicated that existing financial schemes are not fully adequate to meet salinity control programs. This is due either to insufficient magnitude of available funds or lack of legal authorization. Salinity control below Hoover Dam is, however, a possible practical approach toward minimizing the economic impact of salinity and should receive further consideration in the formulation of a basinwide salinity control program. Fluctuations in salinity concentrations result from factors such as seasonal changes in stream flow and water use control throughout the basin.

The water quality simulation model utilized to predict future salinity concentration determines only long-term average concentrations and does not have the capability to predict the magnitude of short-term fluctuations. Water quality simulation capabilities will therefore need to be refined before the effectiveness of control measures can be evaluated.

In the past, development of the basin's water resources was guided primarily by two basic objectives: (1) full development of full water supply allocations to each state by applicable water laws and compacts, and (2) expansion of the regional economy. There is an urgent need for a water quality objective to supplement these basin objectives and provide guidance in optimum development of the remaining water resources. It is possible, it has been demonstrated, that one can control and manage the salinity on an economically practical basis in an economically practical way. In addition, the feasibility of maintaining salinity concentrations at or below the levels in the Colorado River below Hoover Dam has been shown. The enhancement of water quality in the lower river would alleviate much of the future economic impact of salinity. It is recommended that a broad water quality objective be adopted by the basin interests, which would require salinity concentrations to be maintained at or below present levels in the lower Colorado River. This objective would become part of the basic policy guiding the comprehensive planning and development of the basin's remaining water resources.

It is recommended that appropriate Colorado River basin states take the steps necessary to establish numerical objectives for salinity concentrations based on factors such as average concentration of total dissolved solids for any given month to be maintained below 1000 mg L^{-1} at Imperial Dam.

Salinity Control Program by Basinwide Jurisdiction

A single institution with basinwide jurisdiction must be developed. It appears that it would be necessary to create a new institution with the necessary authority to plan and implement a control program. There are three possible ways of creating a salinity control agency. The task group can be assembled to formulate salinity criteria. A second possible approach would be to extend the authority of an existing agency or commission to provide the necessary powers to carry out all the phases of a basinwide salinity control program. Perhaps the most desirable approach would be to create a new, permanent state and federal agency or river basin commission with the authority to carry out all the activities necessary to the basinwide management and control of salinity. The third approach seems to be the most reasonable in view of the magnitude and scope of the salinity problem. Consideration should be given to the possibility of extending the authority of an existing agency or commission to assume this responsibility.

The existing legal and institutional arrangements are not adequate to provide the basis for implementing a large-scale salinity control program. Therefore, it is necessary that congressional authority and funding implement a salinity control program. Due to the scale and type of control projects included in the salt load reduction program, an approach similar to that utilized for the authorization and funding of water resource development is suggested. It is recommended that legislation be introduced in the near future to authorize the entire basinwide salt load reduction program and to appropriate funds for the necessary planning studies.

Proposed Salinity Control Program

The planning phase should be thorough enough that it can provide sufficient information for developing an implementation plan and provide feasible reports on which requests for construction funds for necessary control work can be based and for identifying the direction, operation, and costs that should properly be assigned to basin states and other beneficiaries; for a systematic evaluation of the quality and economic aspects of the salinity problem provided by a key element in the project; and for the determination of the potential feasibility and practicality of a basinwide salinity control program. A systems analysis capability similar to the methodology developed for this evaluation will be required for the planning phase.

The project's water quality simulation model is basically a water and salt budget model, with capabilities to predict long-term averages for stream flow, salt loads, and salinity concentrations at various points in the basin and to evaluate the long-term effects of modification in water use and salt loading at any point in the river system. We must be able to update the model.

The final or implementation phase of the basinwide control program would include the appropriation of construction funds, the actual construction of the projects, and the actual management of salinity through operation of control works. When these control measures are implemented, provision will be needed for funding for continued operation and maintenance that most facilities will need continuously for the foreseeable future.

SUMMARY

The history of Colorado River development is reflected in the law of the river, which has been the ruling force since development started. It has had to evolve to provide not only certainty of flow, but flexibility. Doctrines may have lagged behind changes in public attitude, but they have not prevented water policy changes to serve changing values, which include a consideration of insular minorities and environmental quality. The environmental doctrine of Western water law was the rule "first in time, first in right." This made possible the diversion of water for projects sometimes quite remote from the basin.

The Prior Appropriations Doctrine prevented large upstream diversions that might drain the water supply. The water law was fashioned not only to allocate water and to facilitate development but also to cushion development from the risk of shortages and drought. As water use increased for growing agricultural and urban demands, the Western states persuaded the federal government to secure new supplies to carryover storage and interbasin diversion projects. This was done to increase development of the West. President Theodore Roosevelt envisioned that the western half of the United States would sustain a population greater than that of the entire country of that day if water that now runs to waste were saved and used for irrigation.

The Reclamation Act of 1902 provided loans to attract small farmers to these previously arid lands. According to the Congressional Report of 1914, the Reclamation Extensive Act was specifically intended to create an irrigated empire in the West. The Colorado River Compact of 1922 divided the Colorado River drainage basin into the Upper Basin, comprising the states of Colorado, New Mexico, Utah, and Wyoming, and the states of the Lower Basin: Arizona, California, and Nevada. In addition to providing equity between the Upper and Lower Basins, the major purpose of the contract was to establish the relative importance of different beneficial uses to promote interstate comity, remove causes of controversy, and secure expeditious agricultural and industrial development of the Colorado River basin: that is, storage of its waters and the protection of life and property from floods. The negotiators of the compact assumed that they were dividing an average annual flow at Lees Ferry of 16 million acre-feet, providing 7.5 million acre-feet annually to each basin. Article III(b) of the contract provided to the Lower Basin, grants for an additional right of 1.5 million acre-feet. Article III provided that the Upper Basin states would not cause the flow at Lees Ferry to be depleted by 75 million acre-feet for any consecutive 10 years, thus protecting the Lower Basin. The basin states provided that any amount granted to Mexico by treaty would be divided equally between the Upper and Lower Basins.

Just as the Grand Canyon is a living geological record, the law of the river can be measured by time. Time helps to define the law's various levels of command. The law adopts different time horizons, from the concept of the eternal to the frantic pace of an hour. A time horizon also helps to chart the ever-increasing complex of accelerations that are changing the law. The law of the river includes the totality of compacts: international agreements, treaties, judicial decisions, statutes, water allocation, and administrative practices. The problem of water supply became one of the great problems in the United States and involved President Roosevelt, President Eisenhower, and some of the more recent presidents of the United States.

Diversion of water became the subject of many controversies and has never really been settled. California has made the greatest demands. Recently, California installed subsurface irrigation which has saved about 50% of the water supply (because 50% is lost in evaporation). This has gone a long way toward mitigating the problem.

It was mandated that water of a quality suitable for drinking be delivered to Mexico. The salinity of the water was ever increasing, due to reuse and evaporation. Various methods of desalination have been proposed. The problem has not been solved but has been mitigated.

BIBLIOGRAPHY

Committee to Review the Glen Canyon Environmental Studies. 1991. Colorado River ecology and dam management: proceedings of a symposium, May 24–25, 1990, Santa Fe, N. Mex. National Academy Press, Washington, D.C.

Committee to Review the Glen Canyon Environmental Studies. 1996. River resource management in the Grand Canyon. National Resource Council. National Academy Press, Washington, D.C.

Ingram, H., A. D. Tarlock, and C. R. Oggins. 1991. The law and politics of the operation of Glen Canyon Dam. pp. 10–27. *In*: Committee to Review the Glen Canyon Environmental Studies, (eds.), Colorado River ecology and dam management: proceedings of a symposium; May 24–25, 1990, Santa Fe, N. Mex. National Academy Press, Washington, D.C.

Reisner, M. 1993. Cadillac desert: The American West and its disappearing water. Penguin Books, New York.

U.S. Department of the Interior. 1997. Quality of water. Colorado River Basin Progress Report 18.

Index